Jean-Claude Gall

Ancient Sedimentary Environments and the Habitats of Living Organisms

Introduction to Palaeoecology

With 130 Figures

Springer-Verlag
Berlin Heidelberg New York Tokyo 1983

Professor Jean-Claude Gall, Université Louis Pasteur, 1, rue Blessig, F-67084 Strasbourg

Translator
Dr. Peigi Wallace, Imperial College of Science and Technology, Department of Geology, Royal School of Mines, Prince Consort Road, GB-London SW7 2BP

Translation of the French Edition:
J. C. Gall, Environnement Sédimentaires Anciens et Milieux de Vie. Introduction à la Paléoécologie.
Doin, Editeurs, 8, place de l'Odéon, F-75006 Paris (C) Doin, Paris 1976

Cover picture. Desiccation cracks and reptile footprints (*Cheirotherium*) in relief on a sandstone bedding plane. The animal must have walked across the slab after the puddle had dried up since the footprints cut across the cracks. From a slab of Bunter Sandstone from Hildburghausen (Thuringia) in the collections of the Institute of Geology in Strasburg ($\times$ ¼)

ISBN 3-540-12137-4 Springer-Verlag Berlin Heidelberg New York Tokyo
ISBN 0-387-12137-4 Springer-Verlag New York Heidelberg Berlin Tokyo

Library of Congress Cataloging in Publication Data.
Gall, Jean-Claude, 1936–
Ancient sedimentary environments and the habitats of living organisms.
Translation of: Environnements sédimentaires anciens et milieux de vie.
Includes bibliographies and index.
1. Paleoecology. I. Title.
QE720.G3413   1983   554.4s [560′.45]   82-19697
ISBN 0-387-12137-4 (U.S.)

Reproduction of the figures: Gustav Dreher GmbH, D-Stuttgart
Printing and bookbinding: Konrad Triltsch, Graphischer Betrieb, D-8700 Würzburg
2132/3130-543210

Foreword

I am pleased to be able to introduce this book by Monsieur
Jean-Claude Gall, firstly because it is a book, secondly
because its author has been a colleague for 15 years, and
finally because it is a book which demonstrates the growing
importance of Palaeobiology.

"Because it is a book". I have already commented elsewhere on the value which the Earth Science community
places on a book. And here I am speaking, not of a thesis or
a specialised memoir, which are always precious, but of a
manual or text, which draws on the experts in the service of
all. In the years preceding and following the Second World
War, the number of "books" written by French geologists
could be counted on the fingers of one hand. Today I am
happy to see that the number of geological "books" is increasing in France, taking the word "geology" in its broadest
sense. This I see as a sign of the growth of the Earth Sciences.
Without doubt there are many more geologists, but they have
also become more specialised. Universities, except when
they specialise in studies which are of immediate application,
or when they are held back by financial problems, devote
themselves to their true work: the education of the young
and the pursuit of research. During this time, the geological
surveys have expanded, providing jobs for students and
applied geologists. This is a balanced profession, and specia-
lists on both sides manage to find the incentive and the time
to write books.

"Because Monsieur Jean-Claude Gall is my colleague".
I have held the Chair of Geology at Strasbourg for almost
21 years, and during that time I have lived among first an
increasing, then a stabilising, group of geologists who are
active, inventive and, a particular blessing, very co-operative
and happy. In this environment, teachers, researchers, engi-
neers and technicians have taught me a great deal about many
things. Amongst these, Monsieur Jean-Claude Gall is the
teacher in charge of palaeobiology. It is not that Strasbourg

has a special vocation in this field, but the theoretical and practical teaching, the curation of collections and also the position of palaeobiology in the sedimentological research of our Institute, require a team knowledgeable in this disccipline in our large establishment. I would like to thank Monsieur Jean-Claude Gall here for the excellence of his teaching, for the quiet helpfulness of his service and for his rock-like loyalty.

"Because this book demonstrates the growing importance of Palaeobiology". The fortunes of my career have meant that for 30 years I have been involved in teaching palaeobiology, though my personal researches were involved in sedimentary geochemistry and diagenesis. This means I can be frank. There is a continuous feeling in the Earth Sciences which threatens the respect due to palaeobiology, or even its simple survival. This is not rational, and those geologists who are not palaeobiologists must be aware of this.

Palaeobiology is the foundation of the science of evolution — the original theory of evolution is derived from it. From there, it has permeated all disciplines, including the humanities. Through palaeobiology, we can demonstrate the "rhythms and scale of evolution" through time. This is a major philosphical and scientific question, which is based in the Earth Sciences.

Palaeobiology is one of the main disciplines in which one can learn the laws and lessons of biometry. No study of synchronous biometry, undertaken amongst contemporaneous populations, can neglect diachronous biometry, which has affected evolutionary lineages throughout their development. This is of great present-day interest when we consider the dynamics of populations and of societies.

Palaebiology is the most precise tool which man has for the measurement of time. At the present time, great efforts are being made in the field of absolute geochronology, by means of isotope dating. The results are spectacular and allow us to date the Precambrian rocks of Africa, lava from the moon and the appearance of man in eastern Africa. Nevertheless, palaeobiology is still the most accurate tool we have for identifying relative geochronology. Ammonites, foraminifera, pollen and plankton, rodents and the larger vertebrates are all used to give relative ages not only in stratigraphy, but also in tectonics, petroleum geology and applied geology. In modern plate tectonics theory, without the use of micropalaeontology it could not be proved that the plates move, that they are formed and sink, and travel over younger and younger deposits. To neglect chronological palaeobiology is to refuse to take advantage of the finest chronometer of the history of the earth.

Palaeobiology includes not only the study of extinct organisms but also that of the way they were associated and of the conditions under which they lived. This is where palaeoecology comes in and many palaeobiologists today are palaeoecologists. In the footsteps of Monsieur Jean Piveteau, who wrote the admirable little book on the *Images des mondes disparus,* Monsieur Jean-Claude Gall has here produced a textbook on palaeoecology. Faunal and floral associations exist in equilibrium with their environment. No study of sedimentology, of palaeogeography or of palaeooceanography should ignore ancient environments and their populations, with their story of evolution, of migration and of death. There are thousands of examples in historical geology, in mining and in petroleum geology.

Palaeobiology discusses the organic content of sedimentary rocks, both that which is present and that which, although now disappeared, has affected them. As a man who has spent his life trying to understand sediments and their alteration in terms of inorganic chemistry, I know that this work is only just beginning. The geochemistry and the history of rocks depends not only on the minerals they contain, but also on their organic content. The organic geochemistry of sediments is indispensable: it controls the pH and redox potential of environments, the solubility and combination of certain elements and their migration, their deposition and their accumulation. This biogeochemistry is controlled by the living organisms which inhabited the sedimentary environments and the surface of the continents: palaeobiology is fundamental to modern biogeochemistry.

For these reasons, it is quite clear to me that palaeobiology has always been part of the Earth Sciences and must always remain a part of them. The development of our disciplines can be furthered only by close collaboration between palaeobiologists, structural geologists, geochemists and mining geologists in exploring new fields.

It is in this spirit that Monsieur Jean-Claude Gall has prepared this book. As a man who spans the boundary between palaeobiology and sedimentology, he is as familiar with the

Mississippi delta as with the ancient deltas and shorelines of the Triassic seas of the Vosges. He blends the knowledge of a sedimentologist, the learning of a palaeobiologist, the careful use of his pen and the skills of a draftsman. The description of nine ancient environments, which closes this book, is fascinating reading. I wish good luck to this *Introduction to Palaeoecology.*

1st April 1975

Georges Millot

Preface

One of the main aims of geology is to reconstruct the environments which have followed one another on the surface of the Earth throughout its long history. Such an undertaking involves the description of living organisms and their modes of life as well as the physico-chemical characteristics of their environments. Palaeoecology plays a major part in this field of research. Being the science of the environment, it includes the study of the changing relationships which exist between fossil organisms and the sediment; it calls on disciplines as varied as palaeontology, petrology, sedimentology, geochemistry, etc. At a time when the divisions between disciplines are becoming blurred, palaeoecology appears to be rather an attitude of mind, a way of approaching the subject, than a separate science.

This work sets out to show some of the different ways in which the geologist can study ancient environments. Its aim is to introduce the reader to the methods of deduction used in palaeoecology and to familiarise him with long-gone environments which are often different from those of today.

The information which the fossils and the sediment can give us about ancient sedimentary environments is examined in the first part of the book.

The second part concentrates on palaeoecological syntheses by means of the description of nine reconstructions of continental and marine environments. Apart from Ediacara, which is included because of its importance in the history of life, all the examples are from Europe.

If this conjuring-up of the past succeeds in making historical geology come alive for the reader, the goal which I set myself will be fully realised.

Strasbourg, Easter 1975

Jean-Claude Gall

Table of Contents

Part One
**Information Deduced from the Fossils
and the Sediment**

Chapter 1 Modes of Life

Animals and plants are highly dependent on their immediate environment to satisfy their vital needs for feeding and reproduction. Within populations, natural selection favours the forms best adapted to the physico-chemical and biological conditions which dominate their environment. Thus recognition of the adaptive characters of fossil organisms and an understanding of their significance forms an important part of the reconstruction of fossil environments. This is the aim of **functional morphology**. Broadly, the function of various structures can be reconstructed from the morphology of the hard parts preserved by fossilisation.

A knowledge of the workings and mode of life of present-day organisms is obviously fundamental to the interpretation of fossil species. As in so many aspects of the earth sciences, the present is, to a certain extent, the key to the past.

I. Mobility

How an organism feeds, protects itself against its enemies and reproduces itself is controlled to a great extent by its mobility, and one can base a broad classification of living beings on this (Fig. 1).

1. Aquatic Organisms

a) Benthos

Benthic organisms or benthos live in a very close relationship to the bottom. **Epibiontic** forms live on the surface of the sediment while **endobiontic** forms live within it, either buried or in holes.

i) Sessile Benthos

The adults of sessile species are fixed to, or sit on, the substrate or another organism (epizoans, epiphytes). Since they do not move, they are com-

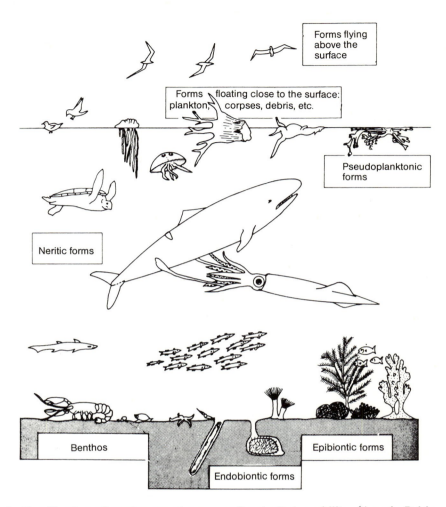

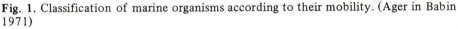

Fig. 1. Classification of marine organisms according to their mobility. (Ager in Babin 1971)

pletely under the influence of their environment. They are thus excellent environmental indicators.

There are many adaptations to a fixed mode of life:

☐ some organisms simply sit on the surface of the unconsolidated sediment (Fig. 2) by means of spines (productid brachiopods) or on their shell (Liassic gryphaeids, whose left valve, in contact with the bottom, is shaped like a cradle to raise the animal above the mud, while the right valve acts as an operculum);

☐ many other organisms are fixed to the substrate (Fig. 3):

— *by a flexible stalk;* for example: most aquatic plants, brachiopods, fixed echinoderms (blastoids, crinoids etc.), some crustacea (goosenecked barnacles) etc;

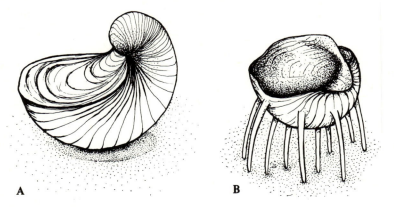

Fig. 2A,B. Sessile benthic forms living on a mobile substrate. **A** bivalve *(Gryphaea arcuata,* Lias)**. B** brachiopod (Upper Palaeozoic productid)

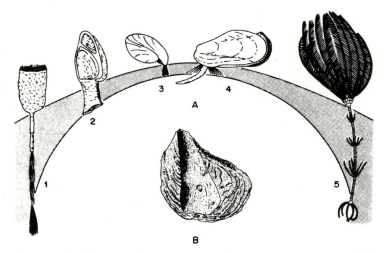

Fig. 3A, B. Methods of fixation of sessile benthos. **A** by a flexible organ: *1* siliceous sponge *(Hyalonema); 2* cirripede crustacean *(Lepas; 3* brachiopod *(Magadina); 4* bivalve *(Modiolus); 5* crinoid *(Cenocrinus).* (Zeigler 1972); **B** by their shell: valve of a Sparnacian *Ostrea uncifera* showing the impression of its attachment area. (Plaziat 1970)

– *by part of their shell or skeleton* which welds itself to the substrate and mimics its shape; for example: sponges, corals, bryozoans, annelids (serpulids), bivalves (oysters, rudistids), crustacea (barnacles), etc.;

– *by a specialised organ;* for example: the exothecal lamellae of archaeocyathids, the byssus of bivalves etc.

In general, a fixed mode of life favours the development of external skeletons, shells and carapaces, as a means of protection against predators.

They frequently have a radial symmetry (sponges, coelenterates, echinoderms), and their distribution takes place by means of free larvae.

ii) Vagile Benthos

The movement of vagile or free species is limited by their contact with the substrate, which effectively controls their search for food. They move in several different ways:
— *by contractions of their body muscles;* for example: worms;
— *by a contractile foot;* for example: many molluscs (gastropods);
— *by appendages for locomotion;* for example: the parapodia of annelid worms, arthropod feet (trilobites, limulids, crustacea);
— *by propulsion organs;* for example: cephalopod jets, fins of benthic fish (skates and rays, Palaeozoic armoured fish), bilaterally symmetrical shells of some bivalves (pectinids), etc.;
— *by specialised structures;* for example: the ambulacral system and spines of echinoderms.

Animals which move actively are generally **bilaterally symmetrical**.

iii) Infauna

The infauna includes organisms which **burrow** into unconsolidated sediments (Fig. 4) or bore into hard substrates (Figs. 17, 36). They are also described as **endobiontic**. They come from several different zoological groups: worms, molluscs, crustacea, echinoderms. This mode of life requires certain morphological adaptations:

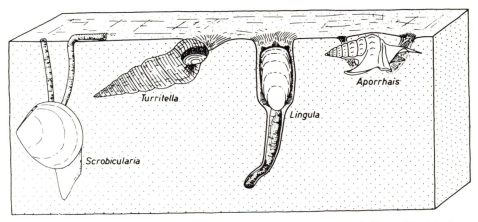

Fig. 4. The endofauna of Recent mobile bottoms: brachiopod *(Lingula)*, bivalve *(Scrobicularia)*, gastropods *(Aporrhais, Turritella)*. (Zeigler 1972)

— *a trend towards the reduction of carapaces and shells* whose protective role has become unnecessary because of the burial of the animal; for example: wood-boring (teredinids) or rock-boring bivalves (pholads);
— *the development of a siphon,* a fleshy, tubular extension of the body of certain molluscs (bivalves, gastropods) which passes water through the pallial cavity. In the bivalves (Fig. 5), the formation of a notch in the pallial line, the **sinus**, shows where it enters the valves. The shell also becomes elongated in an antero-posterior direction and frequently gapes around the siphon;
— *a change in the ambulacral areas* on the upper surface of the test of irregular echinoderms into branchiae (**petaloid ambulacra**) (Fig. 6).

If body fossils are not present, the infauna leaves traces of its activity in the sediment which have great palaeoecological interest (p. 40).

b) Nekton

Nektonic organisms or nekton live in the body of the sea where they move actively in search of food. They move by means of **swimming organs:**
— *the fins of aquatic vertebrates;* in fish and some tetrapods (ichthyosaurs, dolphins, whales) the body is propelled by the caudal fin; elsewhere, this function is carried out by paired structures changed into **paddle-like fins** by the elongation or increase in number of joints (plesiosaurs, turtles, seals) (Figs. 9, 116);
— *the jet and funnels* of cephalopods;
— *the caudal fan and swimming appendages* of decapod crustacea.

The bodies of animals well adapted to swimming often have a spindle-shaped **hydrodynamic profile** (fish, reptiles, mammals). This is a remarkable example of morphological convergence (Fig. 7).

c) Plankton

Planktonic organisms or plankton live free in the sea and are passively swept about by it. Depending on whether they are animal or vegetable, they are known as **zooplankton** or **phytoplankton**. In general, planktonic species do not have organs to help them move. Because of their low body density, they are able to **float**. This low body density can be achieved in several different ways (Fig. 8):
— *by a very small size;* many forms are microscopic; for example: the protista, the larvae of various metazoans, etc.;

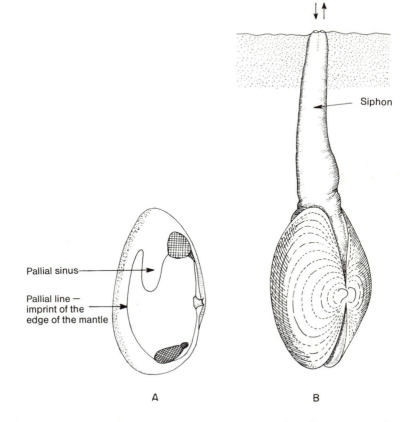

Pallial sinus

Pallial line —
imprint of the
edge of the mantle

Siphon

A   B

Fig. 5A, B. Adaptation to a burrowing mode of life in the bivalves *(Mya arenaria):*
A internal view of the left valve; **B** animal in position of life. (Modified after Boué
and Chanton 1958)

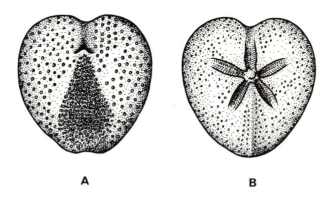

A   **B**

Fig. 6A, B. Adaptation to a burrowing mode of life in the irregular echinoids *(Micraster*
– Upper Cretaceous). **A** jawless oral face; **B** apical face with petaloid ambulacra.
(Devilliers 1973)

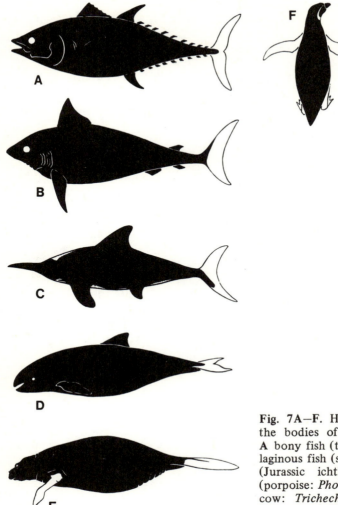

Fig. 7A–F. Hydrodynamic shapes of the bodies of swimming vertebrates. A bony fish (tuna: *Thunnus*); B cartilaginous fish (shark: *Lamna*); C reptile (Jurassic ichthyosaur); D mammal (porpoise: *Phocaena*); E mammal (seacow: *Trichechus*); F bird (penguin: *Spheniscus*). (Gutmann 1966)

– *by the absence* (medusids) *or the reduction of the skeleton* (the perforated shells of radiolaria; the reduced shells of heteropod and pteropod gastropods);
– *by expanding the body,* thus increasing its surface area; for example: the calcareous spines of globigerinids and some coccolithophorids; the appendages of certain acritarchs; the development of appendages and bristles on crustacean larvae, etc.;
– *by the secretion of small droplets of oil;* for example: the coccoid family of green algae which give rise to algal hydrocarbons;
– *by gas-filled floats;* for example: some algae (sargassids), graptolites etc.;
– *by high tissue water retention;* for example: medusids, some tunicates etc.

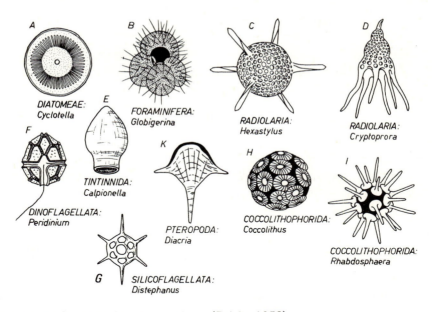

Fig. 8. Recent and fossil planktonic organisms. (Zeigler 1972)

The plankton play an important role in the **food chain** of present-day seas and lakes. Their small size and their reduction in skeletal development means their importance has often been underestimated.

Nekton and plankton are also grouped as **pelagics** since, unlike the benthos, they are independent of water depth.

d) Pseudoplankton

The term pseudoplankton groups together sessile organisms which occasionally fix themselves to **floating objects** (algae, wood, shells, etc.) (Fig. 117). Like the plankton, they are subject to the play of the currents. They come from many groups of organisms: coelenterates (hydrozoans), bryozoans, annelids (serpulids), molluscs (bivalves, gastropods), arthropods (cirripede crustaceans), echinoderms (crinoids), etc. When the floating support is not fossilised, pseudoplanktonic species can easily be confused with benthic forms and this can lead to inaccurate environmental interpretations.

2. Land Organisms

The ability to live on land is one of the great developments in the history of life. During the Palaeozoic, plants and several groups of animals (arthropods, gastropods, vertebrates) in turn colonised the continents. Air breathing and protection against desiccation were among the many adaptations required by this change in environment.

A **fixed mode of life** is found only among autotrophic organisms which take their nourishment directly from the soil. This is so for the plants. In the animals, however, the search for food requires **active travel** on the ground or in the air.

a) Movement on Land

Travel on land can occur in various ways:
- *crawling* by means of contraction of the body muscles (worms, caterpillars) or by the use of a specialised organ such as the foot of the gastropods;
- *walking* with articulated appendages; for example: the arthropods and the vertebrates.

Tetrapod vertebrates, with pentadactyl extremities, provide good examples of adaptation to different modes of travel (Fig. 9):
- *running,* helped by a more erect posture combined with a reduction in the number of digits (horses);
- *climbing* by the acquisition of claws, adhesive pads or by the ability to oppose the digits to make grasping structures (primates, some marsupials);
- *jumping* using an articulated hind limb with three sub-equal segments (frogs, kangaroos, rabbits);
- *burrowing* by shortening and enlarging the fore-limbs (moles).

Elsewhere, among tetrapods which have returned to an aquatic mode of life, the pentadactyl extremity can be changed into **swimming paddles** by an increase in the number of joints or digits (ichthyosaurs, plesiosaurs, cetaceans etc.).

b) Flight

The development of **wings** has occurred in the arthropods (insects) and the vertebrates.

The two pairs of wings in insects are a membraneous expansion of the dorsal side of the thorax. They are lost in parasitic species.

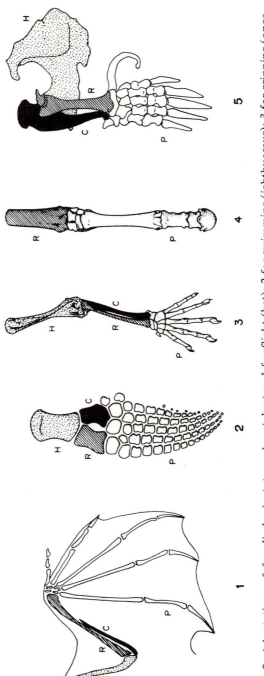

Fig. 9. Adaptations of fore-limbs in tetrapod vertebrates. *1* for flight (bat); *2* for swimming (ichthyosaur); *3* for gripping (opossum); *4* for running (horse); *5* for burrowing (mole). *H* humerus; *R* radius; *C* cubitus; *P* phalanges (finger- and toe-joints). (*1* and *5* from Grasse 1967, *2*, *3* and *4* from Devilliers 1973)

Vertebrate wings are altered fore-limbs (Fig. 9). In flying reptiles (pterosaurs) and bats (cheiropterans), they consist of a web of skin, supported by one or several digits. In contrast, in the birds, the lifting surface of the wing is made of the arms covered in feathers.

II. Nutrition

The search for food controls the distribution and density of organisms.

1. Autotrophic Organisms

Chlorophyll-bearing plants use solar energy to carry out organic synthesis from mineral constituents. Autotrophic forms also exist in the **bacteria**. Some of them can occur in the absence of free oxygen. They are amongst the first signs of life on earth.

All animals are **heterotrophic**. They feed on organic matter of animal or vegetable origin.

2. Microphagous Organisms

Microphagous organisms eat small nutrient particles or small organisms (plankton) either from suspension in the water or mixed with the sediment.

a) Suspension Feeders

Nutrient particles in suspension in the water can be collected in several ways:
- *by beating vibrating cilia* which produce a water current tending to direct nutrients towards the mouth; for example: flagellar cells of sponges, tentacular plumes of some attached polychaetes (serpulids), the lophophore of brachiopods and bryozoans, the food grooves on crinoids' arms, etc.;
- *by filter mechanisms;* for example: the appendages of trilobites and cirripede crustaceans, the sticky filaments of some polychaetes (terebellids), the gills of bivalves (oysters) and tunicates (ascidians), etc.;
- *by organs catching food and carrying it to the mouth;* for example: the tentacles of fixed coelenterates.

The ability to capture larger prey leads to a macrophagous diet.

b) Detritus Feeders

Organic matter often accumulates on the surface of the sediment where it forms a thin film. This is actively gathered by detritus feeders in various ways:
— *grazing* made possible by the development of a mouth (gastropods) or labial palps (*Nucula* — bivalve);
— *by the elongation of the siphon* of burrowing bivalves (tellinids);
— *scuffling* of the sediment with their appendages by many crustaceans (copeopods, amphipods);
— *by the movement of their ambulacral feet* by various ophiuoids.

c) Mud Feeders

Mud feeders eat the organic matter which is disseminated in the sediment by swallowing large quantities of mud or sand. They thus do a considerable amount of re-working of mobile substrates: this is called **bioturbation**.At the same time they produce a large number of **faecal pellets** which become mixed with the sediment. Many annelids do this (arenicolids, lumbricids), as do echinoderms (holothurians, spatangoids), enteropneusts etc.

The microphagous diet is characterised by the **absence of chewable pieces**. It is often associated with a sedentary or barely mobile mode of life.

Several fixed suspension feeders belonging to various groups develop **cone-shaped** shells or skeletons whose opening is orientated towards the water surface. Within especially dense populations, such an orientation allows efficient interception of food particles in suspension. This is a good example of **morphological convergence** (Fig. 10).

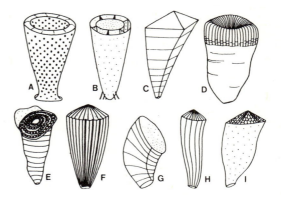

Fig. 10A—I. Cone-shaped tests in various suspension-feeding organisms. **A** Jurassic sponge *(Tremadictyon)*; **B** Cambrian archaeocyathid; **C** Palaeozoic conularid; **D** Jurassic hexacoral *(Montlivaltia)*; **E** Permian brachiopod *(Richthofenia)*; **F** Cretaceous bivalve *(Hippurites)*; **G** Tertiary gastropod *(Rothplezia)*; **H** Tertiary cirripede crustacean *(Pyrgoma)*; **I** Ordovician echinoderms *(Cyathocystis)*. (Zeigler 1963)

3. Macrophagous Organisms

The ingestion of large food particles and the capture of prey are essentially attributes of mobile organisms (Fig. 11).

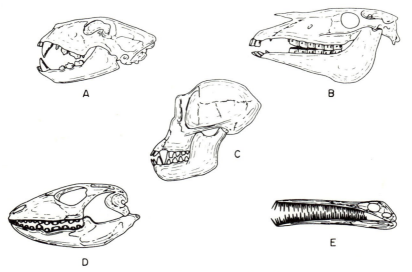

Fig. 11A—E. Adaptation of the dentition of the higher vertebrates to their diet. **A** carnivorous diet (lion: *Leo*); **B** herbivorous diet (horse: *Equus*); **C** omnivorous diet (chimpanzee: *Pan*); **D** dentition for grinding (lizard: *Dracaena*); **E** dentition for filtering (Permo-Carboniferous reptile: *Mesosourus*). (Zeigler 1972)

a) Herbivores

Because herbivorous organisms tear up plants, they need hard parts associated with their mouth. This function is carried out:
— *by the chitinous jaws* of free-moving annelids; they are frequently found separated in sediments: these are scolecodonts;
— *by the radula* of gastropods;
— *by the Aristotle's Lantern* of regular echinoderms;
— *by the jaws and postbuccal appendages* of arthropods (insects, some crustacea);
— *by the continuously-growing teeth* and cutting ridges of mammals (ungulates, rodents).

b) Carnivores

Carnivorous organisms feed on living prey which they pursue or trap. Prey is seized by specialised organs:

— *tentacles armed with stinging (urticant) cells* in the cnidarids;
— *the jaws* of annelids;
— *the radula* of some gastropods which bore into shells (muricids, naticids);
— *the buccal appendages* of arthropods: claws (eurypterids, spiders) and mandibles (crustacea, insects);
— *the arms* of echinoderms (asterids);
— *the tentacles and beak* of cephalopods;
— *the teeth* of vertebrates (fish, amphibians, reptiles, mammals) — the carnivorous mammals are the most specialised.

c) Saprophages

Saprophages feed on dead bodies. This is very difficult to demonstrate in fossils. Examples: worms found in the bodies of insect larvae from the Bunter Sandstone of the Vosges; fungal sclerotes preserved in the rhizomes of Devonian psilophytes from Rhynie (Pl. I Fig. 4).

d) Parasites

In fossils, evidence of parasites can be seen where they have deformed the host (protuberances caused in the body cavity of decapod crustaceans by bopyrids) or where they are found within tissues (nematodes in the cuticle of some Carboniferous scorpions).

III. Reproduction

Methods of reproduction control the ability of an organism to invade an environment and colonise it.

1. Asexual Reproduction

Asexual reproduction occurs in lower forms of life where it allows an exponential development of colonies and in consequence a rapid take-over of the environment (bacteria). In most of the protista (foraminiferids, diatoms), it alternates with sexual reproduction. In some foraminiferids (miliolids, nummulites) the alternation of these two modes of reproduction can be identified by the size of the initial chambers (Fig. 12): the microspheric form produces spores for asexual reproduction, the macrospheric produces gametes.

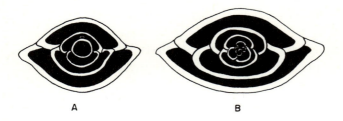

Fig. 12A,B. Alternation of generations in miliolids. A macrospheric form (gamont); B microspheric form (agamont)

A   B

In more highly evolved organisms, asexual reproduction helps in:
— *the dissemination of the species;* for example: plant spores, buds of sponges and hydrozoans, etc.;
— *the formation of colonies* where individuals formed by budding do not detach themselves from their parents; for example: coral colonies, bryozoans, graptolites, tunicates, etc.

2. Sexual Reproduction

Sexual reproduction requires two **gametes** to meet. This can happen in the surrounding water, as is the case with fish, echinoderms and fixed organisms (bivalves, cnidarids, etc.). The pollen of the higher plants is carried towards the ovule by the wind (for example: pollen with little air-filled sacs), water or animals.

On the other hand, among the gastropods, the cephalopods and the majority of species living on land, fertilisation occurs after copulation.

It is possible to distinguish between males and females in fossil organisms when they display **sexual dimorphism**. Apart from the classic example of the horns carried by the males of many ruminants, one can also show sexual dimorphism in brachiopods, bivalves, arthropods, echinoderms and vertebrates. It is particularly clear when the difference between the two sexes is shown by the genital apparatus (insects, eurypterids), the size of the genital opening (some echinoderms), or when there are pouches for incubating the young (ostracods). In molluscs, and especially in the ammonites, the co-existence in some outcrops of two varieties of the same species (formerly placed in different species) differing only in their size and some ornamentation of the adults, has been attributed to sexual dimorphism (Fig. 13).

Eggs generally develop outside the mother (**ovipares**). Many vertebrates (mammals, some fish and reptiles) are **viviparous**: the development of the egg takes place entirely in the genital tract of the mother. Embryos can thus be fossilised in situ (for example: the Liassic ichthyosaurs from Württemberg).

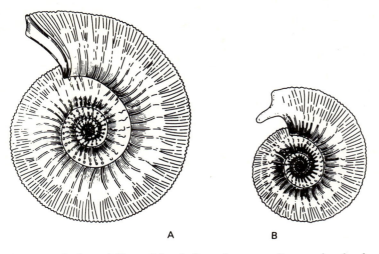

Fig. 13A, B. Sexual dimorphism in Jurassic ammonites previously classified as separate genera. **A** *Cadomites deslongchampsi* (female form?); **B** *Polyplectites linguiferus* (male form?). (Makowski in Babin 1971)

IV. Growth

When one has large populations of a given species available, it is possible to study growth. For many trilobites, one has thus been able to follow the different **larval stages** successively from egg to adult.

The growth of arthropods is discontinuous because it takes place by successive moults. A graph of the distribution of carapace size shows the different moult stages (Fig. 14). In molluscs, one can distinguish juveniles from adults by the density of growth lines on the shell. In many cases, this makes it possible to distinguish between a dwarf population (small sized adults whose growth has been impeded) and a juvenile population (accumulation of young stages).

V. Behaviour

Ethology, the study of the behaviour and habits of fossil organisms, often lends itself to highly imaginative reconstructions, above all where a group is now extinct. Within the last twenty years, research on the mechanical significance of morphological peculiarities and the construction of artificial models has enabled us to understand the modes of life of fossil species much more clearly. This is the case with the feeding habits of the **Tertiary large cats** whose canine teeth developed into sabres (Fig. 15).

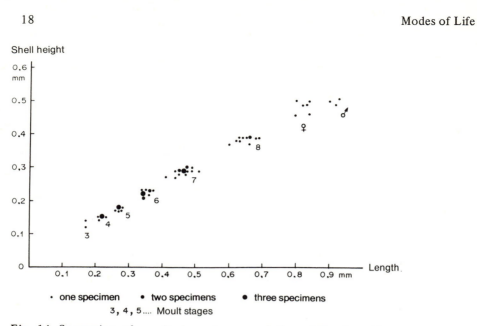

Shell height

Fig. 14. Succession of moult stages in a population of *Hemicyprideis montosa*, an ostracod from the Sannoisian of Cormeilles-en-Parisis. (Keen 1972)

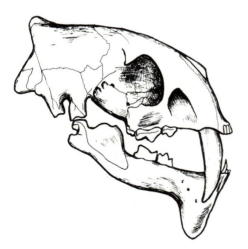

Fig. 15. Skull of *Eusmilus bidentatus* from the Oligocene of Quercy. The upper canines acted like daggers. The anterior of the jaw played the part of the scabbard. (Piveteau 1961)

One can show that the arrangement of their teeth allowed them to stab their prey because of the powerful development of their neck muscles, but prevented them from pulverising the bones. These carnivores could feed only on soft tissue, especially the liver, which they reached by opening the abdomen of their victims.

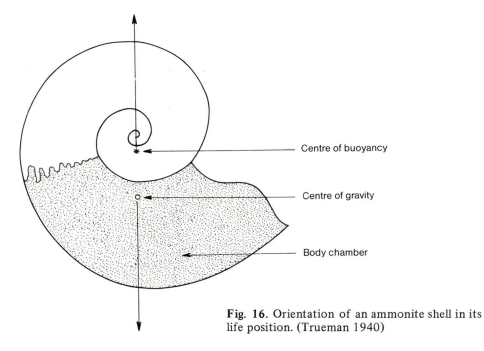

Fig. 16. Orientation of an ammonite shell in its life position. (Trueman 1940)

The **hydrodynamic behaviour** of fossil cephalopod shells has been studied on models where the respective positions of the centre of gravity and the centre of buoyancy were worked out (Fig. 16). Similarly, water circulation in the shells of various brachiopods and molluscs has been studied.

Part Two of this book will provide another opportunity to discuss the habits of many fossil organisms.

Chapter 2 **Constraints on Living Conditions**

Throughout their lives, living beings are subject to the influence of the physico-chemical conditions of their environment. Any change which occurs in the outside world affects organic life; species disappear, others establish themselves. In most cases the forms or populations which occur reflect the conditions obtaining in their environment. Provided one is sure that the fossil animals and plants really lived in the place where they were buried, they are a useful indicator of ancient environments.

Our knowledge of the conditions of life of fossil organisms depends strongly on observations made on their living relatives. This is the principle of **uniformitarianism**, which can, however, lead to dubious extrapolations since the life conditions of many organisms may have changed through geological time without any detectable change in morphology.

Ager cites a particularly eloquent example: in the shallow seas of the Jurassic, the bivalve genera *Pholadomya*, *Trigonia* and *Astarte* occur together in the same outcrops. At the present day, their distributions are very different: *Pholadomya* lives in the deep oceans, *Trigonia* in the warm, shallow water off Australia and *Astarte* is a cold water form.

Thus one must be extremely careful in deducing the life requirements of fossil organisms from their living relatives. Obviously, the older the horizons, the more likely are floras and faunas to differ from present-day populations.

I. Nature of the Substrate

The substrate is the support on which organisms live.

On **land**, it consists of soils, which are rarely fossilised and whose nature depends on both bedrock and climate. The physical aspect of these continental landscapes is demonstrated in the composition of the vegetation and of the herbivorous animals (flora and fauna of the steppes, forests, etc.).

In **aquatic environments**, the texture of the sediment is primarily related to its grain size. Moreover, the establishment of benthic communities is largely controlled by the nature of the bottom. From this, one can determine the nature of the substrate at the moment when it was occupied by the organisms.

There are two categories of substrate: hard bottoms and soft bottoms.

1. Hard Bottoms

The main type of hard substrate is the rocky bottom, though the exoskeleton of other organisms also comes into this category. The heterogeneity of the surface, which causes a certain roughness, favours colonisation by vagile and sessile benthos.

Hardgrounds, which are especially common in Jurassic and Cretaceous carbonates, are formed during a lengthy period of interruption of sedimentation (Fig. 17). This results in an induration of the sediment. These surfaces are often mineralised. They are colonised by **encrusting** organisms (bryozoans, bivalves, serpulids, etc.) and bored into by **lithophagous** animals (sponges, worms, bivalves, cirripedes, crustacea, etc.). These borers penetrate rock and skeletal fragments indiscriminately.

Reefs also establish themselves on hard bottoms.

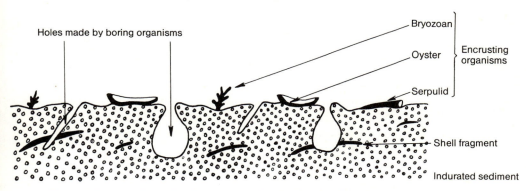

Fig. 17. Section through a hardground in a Bathonian limestone

2. Soft Bottoms

Mobile substrates, sands and muds, have few epibiontic organisms. The **endofauna**, however, is both abundant and varied. The burrows and gallery systems which traverse the sediment turn aside when they come into contact with hard objects, such as pebbles and shells. Thus, when later the sands and muds are indurated into rock, one can distinguish them from borings.

The distribution of burrowing forms is determined by the grain size of the sediments. In argillaceous sediments, the amount of organic matter is extremely important, and favours mud and detritus feeders. Sands provide a more mobile bottom where the water circulation is easy. Their populations are poorer.

II. Salinity

The salinity of water is primarily defined as its **sodium chloride** content. In general, land organisms avoid salt. However, some plants can grow on saline soils **(halophytes)**.

Many aquatic organisms are in osmotic equilibrium with the surrounding water. Other groups have an efficient osmoregulatory system in their organs and tissues. Depending on their ability to tolerate salinity variations, one can distinguish (Fig. 18):
— **stenohaline** organisms which do not tolerate salinity variation;
— **euryhaline** organisms which can tolerate significant variations in salinity.

1. Marine Organisms

Normal seawater (average 35‰ salt) is characterised by **stenohaline** forms. During geological time, whole groups have existed under these conditions: radiolaria, brachiopods, scaphopods, coelenterates (corals), bryozoans, hemichordates, etc. Some stenohaline forms are only known as fossils, including the archaeocythids, trilobites, tentaculitids, graptolites, etc.

The highest faunal diversities occur in marine waters of stable salinity.

2. Brackish Water Organisms

Euryhaline organisms are able to colonise environments whose salinity is markedly lower than that of seawater. Species adapted to these conditions can be found in most parts of the animal kingdom; foraminifera (agglutinated test), bivalves (oysters), gastropods, and the crustacea (ostracodes) are particularly rich in these forms.

Within the Phylum Brachipoda, **lingulids** have inhabited a nearshore environment subject to major salinity variations since the Palaeozoic. Their presence indicates brackish water. Similarly, most **stromatolites** can be attributed to algal structures developed close to the shoreline in water of varying salinity (Fig. 22). The **eurypterids**, giant Palaeozoic

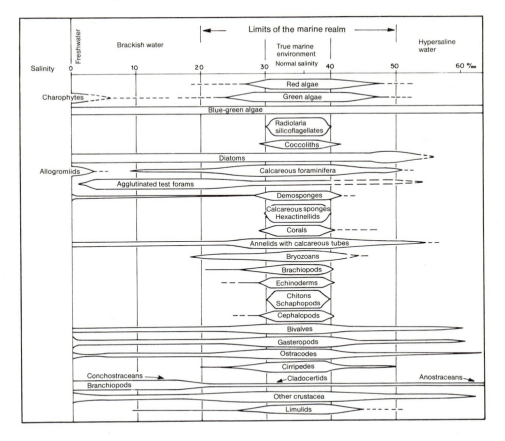

Fig. 18. Distribution of present-day organisms in relation to the salinity of the water. (Heckel 1972)

merostomes, began in the Palaeozoic as marine and gradually adapted to brackish waters in the Silurian and eventually to fresh water in the Carboniferous and Permian.

A list of brackish water species varies considerably from geological period to period, and a detailed stage-by-stage catalogue is far from complete. However, an examination of fossil populations enables us, in most cases, to identify environments of abnormal salinity. Broadly, brackish water faunas are characterised by a **low specific diversity** (Fig. 19). This is due to the instability of the physico-chemical characteristics of the environment which imposes strong selective pressures on aquatic organisms. In contrast, there is usually a **high faunal density**. The size of individuals is also generally rather small compared to members of the same

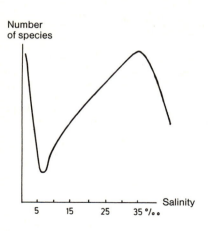

Number
of species

Salinity

5   15   25   35 °/oo

Fig. 19. Diversity of present-day faunas and floras in relation to the salinity of the water. (Remane 1958)

species which lived in water of normal salinity. The shells are also usually thinner than in marine forms.

Similar modifications occur in **hypersaline** waters (salinity greater than 45‰).

The brackish or hypersaline faunas of the Bunter Sandstone and the Keuper of the Germanotype facies compared with those of the Muschelkalk sea illustrate these different adaptations.

3. Freshwater Organisms

Some **stenohaline** organisms are adapted to life in fresh water (salinity less than 0.5‰). Freshwater organisms live in rivers and lakes: annelids, many bivalves and gastropods, crustacea, most aquatic insect larvae, fish, amphibians, etc. The shells of freshwater molluscs are generally thin.

Among the plants, the characians are restricted to fresh or slightly brackish water. Their oogonia, however, which are very resistant, are easily carried by water towards the sea, which reduces their usefulness as environmental indicators.

III. Water Turbulence

The action of the waves and currents which cause **agitation** of the water is of great importance to aquatic populations. It is fundamental to the spread of plankton and larvae. It indirectly controls the establishment of benthic organisms by oxygenating the water and supplying food particles in suspension. In these zones subject to the action of the waves, many

animals protect themselves against mechanical forces by burying themselves in the sediment (suspension feeders) or by fixing themselves to the
substrate. Corals require high energy water. On the other hand, detritus
feeders need quieter waters where food particles can settle out.

The mechanical action of waves and currents can influence the construction of exoskeletons (Figs. 20, 114). Such **adaptations** are known
in red algae, scleractinians, molluscs, bryozoans, etc. In reef environments,
massive corals offering little resistance to the waves are dominant in shallow, turbulent water. In deeper, calmer water, much more fragile, branching forms are developed. From such observations, Lecompte recognised
a bathymetric zonation in the Devonian reefs of the Ardennes. The
Recent gastropod, *Patella vulgata,* also shows a variation in shape depending on the hydrodynamics of its environment. Where it is exposed to the
action of the tides, the shell is conical and thick. It allows the animal to
stick efficiently to the substrate and to retain water in its pallial cavity
during periods of emersion. In pools and areas affected by large waves,
the shell is thinner and more flattened.

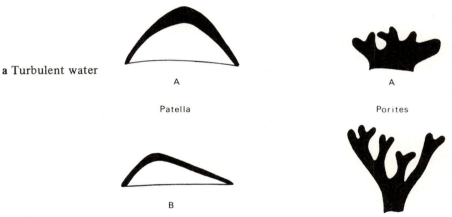

a Turbulent water

A

Patella

A

Porites

B

b Quiet water

B

Fig. 20a, b. Differing shapes of some animals in relation to the turbulence of the water.
(Moore, Vaughan and Wells 1943)

Generally, the orientation, sorting and degree of abrasion of fossils are
good indicators of the degree of energy of the sedimentary environment
(p. 101).

IV. Oxygenation of the Water

In an aquatic environment, free oxygen may have one of two origins. It may come directly from the oxygenated atmosphere by **solution**, through contact with water stirred up by waves and currents. Another source is the **photosynthesis** of chlorophyll-bearing plants (phytoplankton and plant benthos). This depends on the intensity of solar radiation. As this is quickly absorbed by the water, plants can thrive only in water shallower than 200 m — the **photic zone**. In fact, they are only abundant in water depths less than 50 m — the **euphotic zone**. Red algae occur in deeper waters than green algae because of differential absorption of the light spectrum. The symbiotic relationship of scleractinians with brown algae (zooxanthellae) limits the distribution of coral reefs to depths less than 30 m. Below that, these algae cannot photosynthesise.

Because of the double source of oxygen, there is a gradual diminution of oxygen content from the surface waters which are generally saturated, down to very deep water. As a consequence, the fauna thins out. If there is no current to regenerate the deep zones, the water becomes stagnant and a reducing atmosphere with the liberation of **hydrogen sulphide** can develop. These conditions are hostile to life, especially to the benthos and the endofauna. This environment is common at the bottom of the Black Sea, where the fauna is completely absent.

V. Bathymetry

The depth at which aquatic organisms can live is controlled by many factors, especially light and temperature which decrease rapidly as one goes down from the surface waters (Fig. 21).

Chlorophyll-bearing plants and herbivorous animals are restricted to the photic zone (200 m). Below that, one finds carnivores, suspension feeders and detritus feeders. These animals are often blind (for example: deep water ostracodes). The shallow waters of the continental shelf (**neritic zone**) are thus the site of intense biological activity and of a great diversity of forms. At the edge of the sea, the **intertidal zone**, which is subject to the influence of tides, shows many noteworthy ecological characteristics. To protect themselves of the water, animals bury themselves in the sediment, fix themselves against daily exposure and possible temporary freshening to the substrate by suckers or retract into hermetically sealed shells. **Stromatolites**, finely banded calcareous structures which are abundant in the Precambrian and at the present day, form in an analogous environment (Fig. 22). The structures are the result of the trapping of sedimentary particles by a thin layer of blue-green algae.

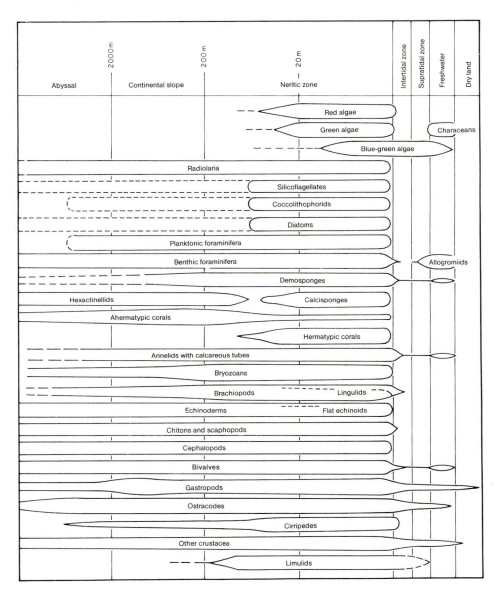

Fig. 21. Distribution of present-day organisms in relation to depth. (Heckel 1972)

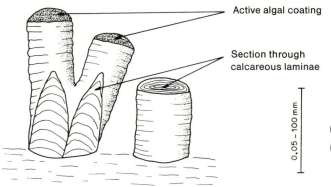

Active algal coating

Section through
calcareous laminae

Fig. 22. Morphology of stromatolites

0.05 – 100 mm

Fig. 23. Internal structure of
an oncolith

Coelenterates	Bivalves, gastropods	Brachiopods
Perisphinctids, cardioceratids	Haploceratids (Oppelids)	Phylloceratids, lytoceratids
Aspidoceratids		

0
100
200
300
400m
Depth

Fig. 24. Bathymetric zonation in the Upper Jurassic of Central Europe. (Zeigler 1972)

They are common in a hot climate. **Oncoliths** are similar, being nodules a few millimetres or a few centimetres in diameter, formed of concentric laminae produced by algae (Fig. 23). However, one can also find such algal structures in freshwater.

Depth zonations, of local value, have been established for several ancient environments. In the *Upper Jurassic of Central Europe,* Zeigler has described the following faunal successions (Fig. 24):

– from 0–20 m, reef forms are dominant; they are associated with algae, brachiopods, molluscs (gastropods, bivalves) and regular echinoids; ammonites are rare;

– from 20–50 m, bivalves thrive; they are accompanied by gastropods and irregular echinoids; ammonites are still rare (perisphinctids);

– from 40–70 m, bivalves are still abundant; ammonites form 20%–30% of the fauna (perisphinctids, aspidoceratids);

– below 80 m, ammonites are dominant (perisphinctids, aspidoceratids, oppelids, phylloceratids, lytoceratids); bivalves and gastropods are rare; brachiopods are still well represented;

– at depths greater than 500 m, ammonites in their turn disappear; all organisms are planktonic (radiolaria).

The remains of **land plants** are not an absolute indicator of the proximity of land because currents can carry them for great distances. Thus coconut fronds have been dredged from the Phillipines trench. Plant roots in situ in the sediment indicate, on the other hand, that any water must be shallow (for example: the *stigmaria* of the Carboniferous forests).

Nevertheless, the bathymetric requirements of many aquatic organisms may have been considerably modified during geological time. At the present day, the **great ocean deeps** are a refuge for many groups which occupied the neritic zone in the past. This is the case, for example, with the hexactinellid sponges, the monoplacophorians, some decapod crustacea (eryonids), attached crinoids, coelocanths and, to a lesser extent, the terebratulids and the cidarids. Some of these seem to have migrated to deeper waters at the time of the late Cretaceous regression.

VI. Turbidity of the Water

To photosynthesise, aquatic plants need light, clear water. Suspension feeding animals also seek these conditions. Water too heavily laden with detritus can clog their food-gathering apparatus and even asphyxiate them. Reef organisms need particularly pure water. However, some forms such as lingulids, starfish and several bivalves easily adapt themselves to a turbid environment and a high rate of sedimentation.

VII. Temperature and Climate

Living organisms are normally only active within a well-defined temperature range. Even when adult forms can tolerate major thermal variations, reproduction and the development of juveniles have much stricter requirements.

1. Land Organisms

On dry land, organisms have to face large fluctuations of temperature and humidity associated with the climate.

a) The Fauna

Only **homoeothermic** (warm blooded) animals, birds and mammals, can maintain a degree of independence of external environmental variations and continue normal activity below 0°C. **Poikilothermic** (cold-blooded) vertebrates flourish only where there is a higher mean temperature. The extiction of the large reptiles of the Mesozoic can doubtless be explained by climatic cooling on all landmasses.

During the **Quaternary**, the alternation of glacial and interglacial phases was marked in Europe by a succession of different mammal faunas (Fig. 25). The warming which corresponds to the last interglacial stage was favourable to the elephant and hippopotamus, while the last glaciation is characterised by reindeer, mammoths and woolly rhinoceros, which were protected from the cold by a thick coat.

b) The Flora

Plants faithfully reflect the local climate. The flora of warm countries is more varied than that of lower latitudes. In a dry climate, plants show **xerophytic** characteristics, thick, hairy cuticle, recessed stomata, etc.), for example: the Triassic flora. On the other hand, the Carboniferous flora, which thrived in the coal swamps, mainly consisted of **hygrophytic** species with thin cuticle and poorly developed roots.

In post-glacial sediments, **palynology** has demonstrated several climatic fluctuations: cold phases are characterised by birch and pine, while hazel, oak and beech thrived in the warm phases.

The reconstruction of climate from the flora must take account of the fact that both altitude and latitude can have a similar effect on the distribution of plants.

Fig. 25. Succession of mammalian faunas during the climatic oscillations of the Quaternary. (Thenius and Kuhn-Schnyder in Theobald 1972)
Beginning of the Quaternary (*1* to *8*). Warm fauna of the mid Quaternary (*9* to *13*). Cold fauna of the late Quaternary (*14* to *17*). *1 Dolichopithecus; 2 Epimachairodus; 3 Tapirus; 4 Hipparion; 5 Mastodon (Anancus) arvernensis; 6 Leptobos; 7 Allohippus; 8 Archidiskodon (Elephas) meridionalis;* 9 Rhinoceros from Merck; *10 Elephas antiquus; 11* Hippopotamus; *12* Macaque; *13* Buffalo; *14* Woolly rhinoceros; *15* Mammoth; *16* Reindeer; *17* Musk ox

2. Aquatic Organisms

Thermal variations are less marked in an aquatic environment than on land.

a) Warm Water Organisms

A raised temperature favours the precipitation of **calcium carbonate** and its fixing by organisms. Shells and carapaces are thick and ornamented. These characteristics are particularly striking amongst organisms which live near reefs (for example: the brachiopods *Uncites* and *Stringocephalus* and the bivalve *Eumegalodon* of Devonian reefs).

Present-day reefs only develop in water whose temperature remains above 18°C. Fossil reefs built by hexacorals can reasonably be assumed to have been constructed under similar conditions. In the French Jura, their distribution in space and time during the late Jurassic was previously interpreted as the result of the migration of a reef zone from the NW to the SE following climatic cooling (Fig. 26). In fact, bathymetric variation caused by a flexing of the continental shelf can easily account for the distribution of these reefs.

Similarly, rudists and Palaeozoic corals must have flourished in warm water.

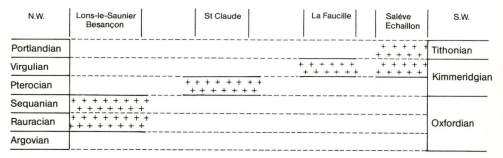

Fig. 26. Distribution in space and time of coralline facies in the Upper Jurassic of the French Jura. (Gignoux 1936)

b) Cold Water Organisms

The low concentration of dissolved lime in cold waters is hardly conducive to the formation of thick exoskeletons. Low latitude bivalves generally have small, thin shells. The faunas are less varied and true reefs are absent. **Siliceous** organisms become proportionately more abundant (diatoms, radiolaria, etc.).

3. Palaeotemperature Measurements

Urey has shown that the relative abundance of the **isotopes O_{16} and O_{18} in carbonates** varies with the temperature of the water in the environment of deposition. The amount of this element in the hard parts of different fossils allows us to calculate the absolute temperature of the water where the organisms lived. Thus the Upper Jurassic belemnites of Scotland show that the temperature of the sea at that time was between 15°C and 20°C, that the animal probably lived through four winters (the curve has four minima) and that the juveniles lived in warmer water than the older forms (Fig. 27). This last observation probably implies that the young belemnites lived in shallower (thus warmer) water than the pelagic adults.

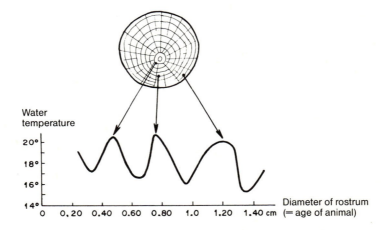

Fig. 27. Palaeotemperatures of the Upper Jurassic sea in Scotland as deduced from O_{16}/O_{18} ratios in a belemnite rostrum. (Modified from Urey et al. 1951)

4. Seasonal Cyclicity

The growth of present-day scleractinians follows diurnal and annual rhythms. These variations are indicated by fine striations on the surface of the calyces. Between two annual events, one can count about 360 daily growth rings. A similar count on Devonian tetracorals gives higher figures, of the order of 400 rings. Thus, in the Devonian, the year must have been about 400 days long. These results can be explained by the progressive slowing of the rotation of the earth about its axis.

Similarly, the succession of the seasons is reflected in the **annual growth rings** of living trees (spring wood and autumn wood). A similar zonation is known in fossil plants as far back as the Devonian. This indicates that seasonal climates have been established at least since that time.

The identification of the different factors which influence the distribution of organisms is purely arbitrary. One single factor is never dominant in an environment. Moreover, interactions occur. For example, temperature influences the oxygen content and the salinity of water and is, in its turn, determined by depth. In order to reconstruct ancient environments, it is extremly important to be able to determine their physico-chemical characteristics and, through them, to explain the diversity of habitats and their populations.

Chapter 3 **Evidence of Biological Activity**

The evidence of biological activity includes remains, other than body fossils, which have been left in sediments by living organisms. They reflect different aspects of life: reproduction (spores, pollen, eggs), feeding (bite marks, coprolites), movement (tracks and trails), habitat (borrows), etc. The difficulties of interpreting this sort of palaeontological evidence are unique: eggs, coprolites and trails are very rarely found associated with their maker. In the majority of cases, the makers are unknown.

This evidence of biological activity reflects the **behaviour** of organisms. In so far as this is a response to environmental conditions, it can provide information about the environment itself. Moreover, these traces are of even greater interest since they often provide the only evidence of life in sediments which are lacking in body fossils. Also, since their fragility means they cannot be transported, they are a guarantee that the organisms were living in the environment in which they were preserved.

I. Evidence of Reproductive Activity

1. Spores and Pollen

Spores and pollen, the study of which forms part of **palynology**, have mainly been used for the reconstruction of fossil floras and climates, especially in the Tertiary and Quaternary. However, because of their small size and resistance, they are easily transported by wind and water for long distances. Thus the information which they can give about the vegetation of ancient environments must be used carefully.

2. Eggs and Clutches

When the eggs of oviparous organisms are protected against impact and desiccation by a resistant envelope, they have fossilisation potential.

a) Invertebrate Eggs and Clutches

For over 100 years, small spherical objects, rich in phosphates and organic matter, which are abundant in Palaeozoic rocks, have been identified as **trilobite** eggs. Similarly, Ordovician chitinozoans have been interpreted as **cephalopod spawn**.

Estherians (branchiopod crustacea) from the Carboniferous, Triassic and Tertiary have yielded egg clusters from inside their carapaces (Fig. 28). The whole mass would be easily dispersed by the wind, which explains the abundance of estheriids in discontinuous environments such as pools and ponds. In the Lower Triassic of the Vosges, the occurrence of two types of egg suggests an **alternation of generations** between clutches which develop immediately (small, numerous eggs), which assure the continuity of the animals in the same environment, and longer-lasting eggs (fewer and larger in size), which represent resistant forms for the dissemination of the species.

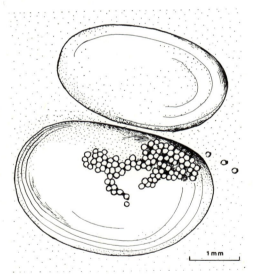

Fig. 28. Shell of conchostracan crustacean (estheriid) from the Bunter Sandstone of the Vosges containing eggs

Insect eggs (Fig. 109) have been described from the same horizon in the Bunter of the Vosges. The eggs, which have an average diameter of 0.25 mm, are encased in a chitinous shell which opens along a median slit (Fig. 29). The clutches contain up to 3000 eggs joined together by **mucilage**. Depending on genus, these are laid out like beads on a necklace or stuck together in a lump like those of present day chironomids. It is possible to work out the internal structures and their relation to the

Fig. 29. Insect eggs from the Bunter
Sandstone of the Vosges

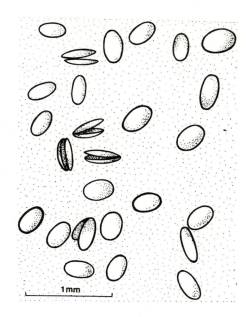

embryo. The sheath of mucilage which surrounds the egg acts as a protection against the desiccation produced during the dry phases of the
Triassic climate.

b) Vertebrate Eggs and Clutches

i) Fish

Eggs of cartilaginous fish, selacians and chimerans have been described
from the Carboniferous and the Triassic. They were protected by a horny
shell which was often extended by filaments used for fixation (Fig. 30).

ii) Reptiles

Large eggs (up to 20 cm diameter), often clustered in nests, have been
found in the Upper Cretaceous of France (Basse Provence, Languedoc,
Corbières), China, Russia, etc. (Fig. 31). They have been attributed to
dinosaurs which laid their clutches close to the banks of lakes or rivers.
These eggs are usually unhatched. A Senonian example from the Gobi
Desert even shows the remains of an embryo. In the South of France (in
the Hautes-Roques region at the foot of Mt. Saint Victoire) Dughi and
Sirugue have shown that shells from the uppermost Cretaceous show
interruptions in growth. According to them, these breaks, which interrupted the formation of the shell walls, are due to abrupt cold spells,

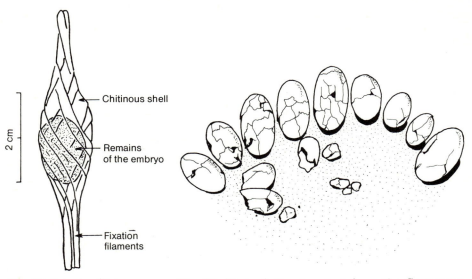

Fig. 30. Selacian *(Palaeoxy-ris)* egg from the Carboniferous of England. (Pruvost 1930)

Fig. 31. Nest of dinosaur eggs from the Cretaceous of the Gobi Desert (after an American Museum of Natural History photograph). Each egg is about 10 cm long

which temporarily interrupted the secretory activity of the oviduct. In these climatic disturbances, they see one of the causes of the extinction of the large reptiles at the end of the Mesozoic.

iii) Birds

Birds' eggs can be distinguished from reptiles' eggs by the microstructure of the shell. They are common in several horizons of the Tertiary.

c) Reasons for Studying Eggs and Clutches

Insect, reptile and bird's eggs are generally laid close to the shore. There a high rate of sedimentation was necessary to bury them rapidly and generally prevent them from hatching.

II. Evidence of Feeding

1. Signs of Predation

It is possible to see wear and breakages which have healed during the lifetime of the animal on the surface of shells, carapaces and skeletons of fossil organisms. In most cases, they can be attributed to the actions of predators. Perfectly circular **holes**, occurring in various shells, are made

Fig. 32. Bivalve shell bored by a carnivorous gastropod *(Natica)*

by carnivorous gastropods (naticids, muricids, etc.) which perforate the shells before ingesting the soft parts (Fig. 32). One can see marks left by echinoid jaws or fish teeth on Jurassic belemnite guards. Bite marks from the teeth of carnivores have been described from many Tertiary and Quaternary mammals.

2. Fossil Excrement

Invertebrate **faecal pellets** form a considerable part of the mobile sediments of the neritic zone. In ancient rocks, they appear as spherical or oval bodies with a diameter less than 5 mm. Their surface often shows constrictions or longitudinal grooves. Elongated faecal pellets with longitudinal internal canals, common in the Mesozoic and Tertiary, are attributed to decapod crustaceans (anomours) (Fig. 33). The organic matter in faecal pellets may be replaced by **glauconite, pyrite** or **phosphates**.

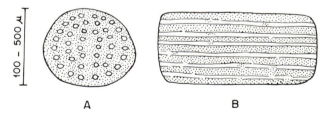

A   B

Fig. 33A, B. Decapod crustacean faecal pellet from the Trias of the mid Prealps. **A** transverse section; **B** longitudinal section. (Brönnimann et al. 1972)

Large specimens of fossil excrement or **coprolites** (several centimetres long) are attributed to vertebrates. They are characterised by an abnormal concentration of bone fragments or broken shell and by a high level of organic matter or phosphates.

The coprolites of certain **fish** have a twisted shape which has been imposed on them by the spiral valvule of the intestine. Larger-sized examples have been attributed to the reptiles (Fig. 34). **Hyena** excrement lends itself particularly well to fossilisation because it is rich in calcium derived from bones ground up by these animals.

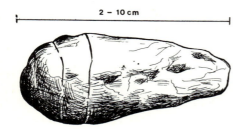

2 – 10 cm

Fig. 34. Reptile coprolite from the Lias of England. (Buckland in Häntschel et al. 1968)

Special mention should be made of **regurgitated pellets** from nocturnal birds of prey, which are frequent in Quaternary cave deposits. They provide important information about the micromammal (rodent) fauna which was the prey of these birds.

III. Trails and Burrows (Ichnology)

Ichnology describes and interprets traces left in sediment by animal activity. These traces or **ichnofossils** (*Lebensspuren* of German authors) include trails, burrows, mines, etc. They are described as **exogenous** or **endogenous** according to whether they are produced at the surface or within the sediment. Many animal traces are produced as natural moulds at the interface between two beds of differing lithologies (for example: sand-clay): thus they form **epireliefs** on the upper side of the beds, **hyporeliefs** on their lower side (Fig. 35).

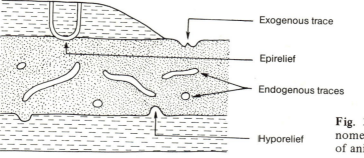

Exogenous trace

Epirelief

Endogenous traces

Hyporelief

Fig. 35. Position and nomenclature of traces of animal activity

Ichnofossils are chiefly the work of **vagile benthos**, more rarely of **terrestrial animals**. It is only exceptionally that they are found with the animals that made them. Similar traces can be made by animals belonging to different zoological groups. Equally, one animal can produce very different traces depending on the nature of its activity (locomotion, feeding, digging, etc.). Thus the classification of ichnofossils is based only on the behaviour of organisms, corresponding, to a certain extent, to the demands of the environment.

1. Dwelling Traces

Many **suspension-feeding** organisms build **burrows** which they inhabit permanently. There they find protection against predators, and, on occasion, against temporary drying-out of the environment. Their occupants collect nutrient particles which pass close to the openings of the burrows. These have to be frequently renewed because of water action. This is why dwelling traces are common in shallow water environments, especially in the intertidal zone.

In the high energy environment of the intertidal zone, hard substrates are bored into by **lithophagous** organisms (sponges, annelids, hydrozoans, bivalves, crustacea, echinoderms, etc.) (Fig. 36). On mobile substrates, the construction and maintenance of burrows requires a low level of water movement. One can identify two types of habitat:

a) **simple burrows** in the form of a straight tube or pocket. The organisms maintain contact with the surface of the sediment by means of siphons (bivalves) or by the action of a retractile pedicle (lingulids). Arthropods (crustacea, some trilobites) protrude their anterior appendages from the burrow, while sea urchins circulate the water by the movement of their tube feet.

b) **U-burrows** connect to the surface by two openings (Fig. 37). In such dwelling, water circulation can be maintained by contractions of the animal's body (annelids) or by the movement of appendages (arthropods).

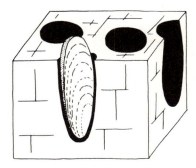

Fig. 36. Dwelling cavities of a lithophagous bivalve *(Lithophaga).* (Zeigler 1972)

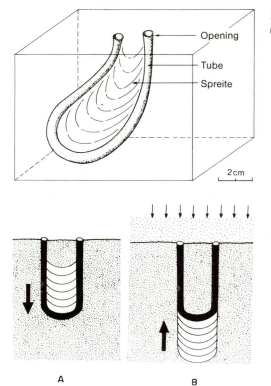

Fig. 37. Crustacean dwelling burrows *(Rhizocorallium).* (Gall 1971)

Fig. 38A, B. Development of a U-burrow as a result of the rate of sedimentation. **A** normal growth; **B** fight against burial in a turbid environment. (Modified from Seilacher 1967)

When the rate of sedimentation is low, the growth of the organisms requires an enlargement of the burrow. If this occurs at depth, successive growth stages can be seen in the sediment as **spreite** spread between the internal walls of the two vertical branches of the burrow. On the other hand, when the rate of sedimentation is high, the animal reacts against burial by continuously raising its burrow: the spreite thus occurs on the outside of the burrow (Fig. 38). U-burrows are made by various animals: annelids *(Arenicola, Polydora),* crustacea *(Corophium),* insects (ephemerid larvae), etc. *Rhizocorallium* is a spreite burrow known from the Cambrian to the Tertiary: the scratches on the walls of the tubes suggest it is made by a crustacean (Pl. I, Fig. 3).

2. Moving and Resting Traces

Vagile organisms create tracks in the sediment which record their movement or their resting marks.

a) Moving Traces

Arthropods and tetrapod vertebrates leave the marks of their appendages or feet on the surface of an unconsolidated sediment (Figs. 39, 118). The resulting trace is generally preserved in relief on the base of the overlying bed. When the sediment is finely bedded, the prints can be transmitted from one lamina to another.

In certain cases, the arrangement of footprints enables one to identify the maker and the length of its stride. For *Iguanadon* (Cretaceous dinosaur), one can recognise running, walking and resting tracks.

Movement traces are good **depth indicators**: reptile and bird footprints are formed beneath a thin layer of water (Fig. 40); limulid tracks are limited to intertidal waters (Fig. 41).

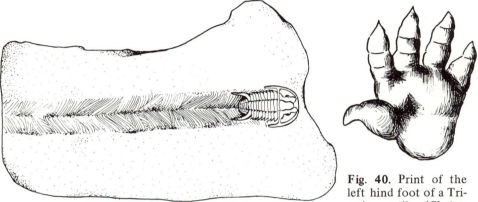

Fig. 39. Trilobite movement trace *(Cruziana)*

Fig. 40. Print of the left hind foot of a Triassic reptile *(Cheirotherium)*

Direction of movement → — Print of fifth appendage

← — Print of telson

← — Prints of anterior appendages

Fig. 41. Limulid trail

Most fossilised invertebrate traces are made by animals moving across bedding: the passage of digging animals (annelids, gastropods, arthropods, echinoderms, etc.), escape traces of animals accidentally buried by an influx of sediment (Fig. 42), etc. Striations, furrows and constrictions occur on their surface; they are the result of the activities of their maker and are used to identify the different ichnofossils.

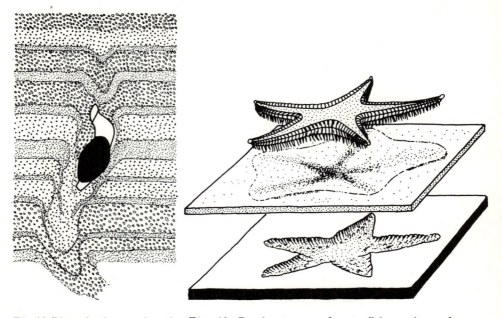

Fig.42. Bioturbation produced in sediment by a bivalve *(Mya).* (Schäfer 1963)

Fig. 43. Resting traces of a starfish on the surface and within the sediment *(Asteriacites).* (Seilacher 1954)

b) Resting Traces

To hide from predators, vagile benthos can temporarily bury itself in the sediment. This results in resting traces which more or less faithfully mimic the shape of the animal's body (trilobites, crustacea, ophiuroids, fish, etc.) (Fig. 43). When the organism's position has been determined by a current, these traces show a preferred orientation (Fig. 91a). Resting traces are confined to the photic zone (0–120 m). Below that, permanent darkness is enough to conceal animals from their enemies.

3. Feeding Traces

a) **Grazing traces** with a sinuous pattern (helminthoids) are made by **detritus-feeding** organisms which methodically exploit the thin organic-rich layer which carpets the mud and bedding planes (Fig. 45).

   b) **Sediment feeders** make a **complex gallery system** progressively infilled by the movement of the animal. Their path can be easily distinguished by the difference between the matrix and the sediment infill. Some of these systems are made by the extension of an initial gallery

Fig. 44. Gallery system of a mud-eating organism *(Chondrites)*. (Simpson 1957)

(Chondrites) (Fig. 44); their regularity caused early authors to mistake them for plant traces *(Fucoides)*. Other gallery systems develop laterally or vertically from a U-burrow *(Gyrophyllites)* (Fig. 46).

Feeding traces are abundant in quiet water deposits where fine-grained sediments are deposited. They are more typical of deeper water environments than those characterised by dwelling traces.

4. Reasons for Studying Trace Fossils

Trace fossils provide a great deal of information about ancient environments.

a) Bathymetry

Tetrapod vertebrate tracks unambiguously indicate the existence of a thin layer of water at the surface. In another connection, Seilacher has shown that the form and function of burrows and burrow systems change with depth (Fig. 45):

- **simple burrows,** straight or U-shaped, arranged approximately perpendicular to the surface of the sediment, are dominant in shallow water; they are the habitat of suspension-feeding organisms which protect themselves from predators, desiccation or desalination by burying themselves. **Resting traces** reflect the same need for protection in a shallow water environment;
- the construction of a **complex system of galleries,** either oblique or horizontal, is the work of deposit-feeding organisms seeking their food within the sediment. Deposition of fine organic particles can take place only in quiet water, in other words, in depths generally greater than those characterised by dwelling burrows;

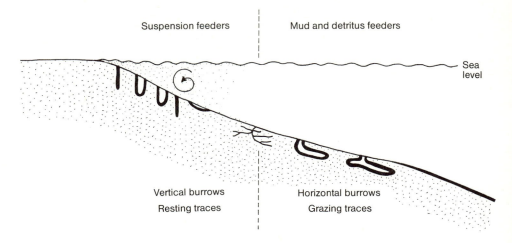

Fig. 45. Bathymetric distribution of the traces of animal activity. (Modified from Seilacher 1967)

— **grazing traces** reflect the exploitation of the thin film of organic matter occurring on the surface of the sediment or at bedding planes; they are abundant in deep water deposits laid down by turbidity currents (flysch facies) (Fig. 46).

b) Oxygenation of the Environment

With some exceptions, benthic and endofaunal organisms need normally aerated water and sediment to flourish. Anaerobic bottoms are generally without life; for example, the Black Sea. The absence of trace fossils from a sedimentary horizon should be regarded as an indicator of an insufficiently oxygenated environment.

c) Sedimentation Rate

The special structure of some burrows (for example, some forms of *Rhizocorallium*) is the result of a flight reaction in order to escape asphyxiation by rapid sedimentation (Fig. 38). Conversely, surfaces scraped by lithophagous organisms correspond to hard bottoms which did not receive any sedimentation for a prolonged period.

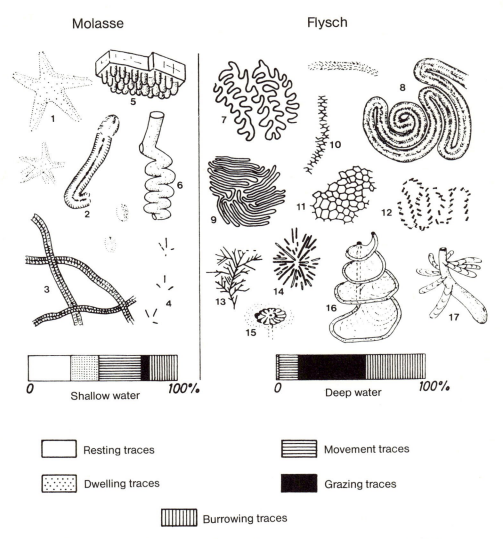

Fig. 46. Traces of animal activity as depth indicators in molasse and flysch. (Seilacher in Zeigler 1963)

d) Cohesion of the Substrate

The presence of the borings of lithophagous organisms or the galleries of deposit feeders indicates the state of induration of the substrate during occupation by the organisms (Fig. 17).

Except in a limited regional context, ichnofossils are not indicative of **Palaeosalinity**.

Chapter 4   The Sediment

By studying sediments, we can learn about the origin of the material, the conditions of its transport and the characteristics of its environment of deposition. Observations can be made at different levels:
− at individual particle level;
− at bed level;
− at outcrop level.

I. Observations on Sedimentary Particles: Sedimentary Petrology

1. Petrological Characteristics of the Sediment

a) Information on Regions Upstream of the Site of Deposition

i) Source Areas

Detrital sediments are derived from areas undergoing erosion, whether nearby or distant. Their mineralogical composition, and above all the presence of pebbles which have survived transport, can tell us about the geological nature of the source area. It is often possible to locate this area by means of present-day outcrops or boreholes.

Thus it has been established that the Bunter Sandstone of the Vosges is derived from uplands situated in the position of the present Paris Basin. A coarse horizon, the Conglomerat principal, has yielded lydite pebbles with Silurian graptolites and quartzite pebbles containing Devonian spiriferids. Recent boreholes have penetrated Silurian and Devonian sediments under the Mesozoic of the Paris Basin. These results confirm those previously obtained from palaeocurrent directions (Fig. 55).

Source areas can also be found from a study of **heavy minerals**: staurolite, kyanite, andalucite, garnet, tourmaline, zircon, etc. These minerals characterise different metamorphic zones in the basement. In the Aquitaine Basin, Vatan was able to trace the source of the sediments which, during the Cretaceous and Tertiary, came partly from the Pyrenees and partly from the Massif Central.

More recently, **thermoluminescence** studies of quartz grains has allowed the identification of different varieties of this mineral and tied them in to their source area.

ii) Climate

Rocks alter at the surface. The physico-chemical processes which cause this alteration are temperature and humidity, which are largely controlled by climate.

In a temperate climate, crystalline rocks weather to sands, where the dissociated minerals, often partially altered, occur in a sparse argillaceous matrix. Quartz remains more or less unaltered. After induration, the sands become **arkoses** (Fig. 47).

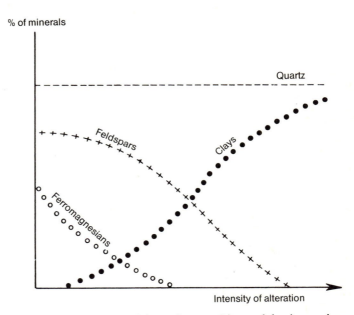

Fig. 47. Change in the mineralogical composition of a granitic sand by increasing alteration in a temperate climate. (Weller 1960)

On the other hand, in a hot, moist **tropical climate**, the rocks are affected by extreme leaching. Silica is partially removed in solution. Aluminium is fixed as gibbsite or combines with the remaining silica to give kaolinite. This results in **laterite** deposits, of which the geological equivalent is the **sideritic** facies associated with outcrops of iron and bauxite.

All minerals, however, do not alter at the same speed: in both temperate and tropical climates, biotite is more vulnerable than muscovite and

and plagioclase alters more quickly than alkali felspar. In arid terrains, plagioclase stays fresher than alkali felspar. From this, one can deduce that the felspathic grits of the Trias, which contain altered felspars, were deposited in a hot climate, alternately dry and moist, intermediate between the temperate zone and the tropics.

b) Information About the Environment of Deposition

i) Redox Potential

Some authigenic iron minerals act as oxidising/reducing indicators. Ferric oxides such as **goethite** [FeO(OH)] indicate an oxidising environment. **Siderite** (CO_3 Fe) and **pyrite** (FeS_2), on the other hand, are characteristic of reducing environments. These also favour the preservation of **organic matter** and thus the preservation of the soft parts of organisms.

ii) Salinity

Some minerals, such as the ferruginous clays **glauconite** and **berthierine (chamosite)** are formed only in marine sediments. Most occurrences of **phosphate** are also marine. **Evaporites** — rock salt, sylvite, gypsum — indicate a hypersaline environment, but they are very mobile and migrate easily. The resulting cavities in the rock can be filled with other minerals, forming **pseudomorphs** (Fig. 66).

So-called primary **dolomites**, characterised by fine grain size and regular bedding, also seem to originate in highly saline environments with a tendency to the evaporitic.

iii) Information from Clay Minerals

When the deposits are primarily detrital, clay minerals are **derived** from the source areas. Some such as **illite** and **chlorite** come directly from massifs undergoing erosion. On the other hand, **kaolinite** and many **montmorillonites** come from peneplaned landmasses covered by soils rich in minerals.

When deposits are more predominantly chemical in origin, clays can **change** or **re-form** in the sedimentary environment itself. This one finds characteristically marine **glauconites** and **magnesium-rich clays** (saponite, stevensite, attapulgite, sepiolite) characteristic of geochemical environments which are supersaturated in silica and cations by evaporation.

Finally, **diagenesis** alters clays and evidence of their detrital or sedimentary origin can be obliterated.

iv) Information from Geochemistry

Clay minerals are able to incorporate certain elements present in the water in their depositional environment into their crystal structure. These elements, generally as **traces**, can be identified and measured by geochemical analysis. It is thus possible to compare variations in their content in different environments. In seawater, for example, **boron** levels are higher than in freshwater. In marine illitic clays, boron is more abundant (100–300 g/T) than in freshwater sediments (30–100 g/T). In a hypersaline environment, the levels are even higher. Boron can thus be used as a **palaeosalinity** indicator. Other elements such as chromium, strontium and vanadium behave similarly.

c) Information on Diagenetic Changes

Petrological studies can reveal geochemical changes in the sediment after its deposition; grain-growth by iron-rich solutions, concretion formation etc. Diagenesis is responsible for the final appearance of the sediment.

2. Petrography

In petrography one studies the form and appearance of the surface of grains and pebbles. From these one can determine the wear caused by and the nature of the transporting agent.

a) Surface Appearance of Quartz Grains

Observations are carried out on dry grains with a diameter greater than 0.5 mm. Three categories of grain can be identified (Fig. 48):
— *unworn grains:* little or no transport;
— *grains with a dull sheen:* water transport;
— *round, matt grains:* wind transport.

When grains with a dull sheen form more than 30% of a sand or grit, they were abraded in a marine environment. When the proportion is lower than 20%, it is not possible to say whether they were transported in the sea or in a river.

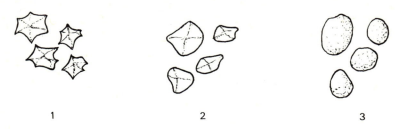

Fig. 48. Different appearances of sand grains. *1* Unworn grains (low transport); *2* Grains with dull sheen (water transport); *3* Round matt grains (wind transport)

Recently, the scanning electron microscope has enabled us to reconstruct the history of quartz grains, thanks to the ability to examine grain morphology in detail.

b) Shape and Roundness of Pebbles

During the course of their transport, particles are subjected to many impacts which tend to produce a spherical shape, although this is rarely completely achieved. The shape of pebbles depends both on their petrography and on the abrading agent.

In the **marine environment**, the pebbles have a rather dull surface. They are more strongly flattened than in a fluviatile environment. Moreover, on beaches their surface often shows circular shock marks (chatter marks).

Wind laden with particles shapes pebbles into **Windkanter**: their surfaces have planar faces separated by crests. A three-sided pebble is called a **Dreikanter** (Fig. 49).

Finally, **glacial** transport causes **striations** on pebbles, which usually have a broadly pentagonal outline (Fig. 50). The striations are the result of stones rubbing against one another within the ice.

Fig. 49. Wind-faceted pebble (Windkanter)

Fig. 50. Glacially striated pebble

3. Grain Size Measurement

The size of sedimentary particles can be measured, either directly (pebbles) or by sieves (sands). For fine-grained sediments (less than 0.05 mm), the settling time can be measured; it is effectively proportional to the grain size.

a) Detrital Rocks

The mean size of detrital particles and the spread of grain size classes depend chiefly on the nature and effectiveness of the means of transport. **Graphs** (frequency curves, cumulative curves) demonstrate the results visually. From the one can calculate **coefficients** for comparison between facies (Fig. 51).

On frequency curves, one can see:
— *the degree of sorting of the grains* from the width of the curve;
— *whether the origin is monogenetic or polygenetic* from the number of maxima shown by the curve;
— *the absence of a fraction of the material* indicated by an asymmetric curve.

Frequency

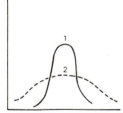
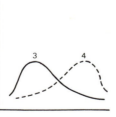

1 Well-sorted sediment
2 Poorly sorted sediment

3 Positive skew
4 Negative skew

5 Bimodal curve
(Polygenetic sediment)

Cumulative frequency (%)

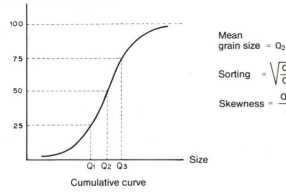

Mean grain size $= Q_2$

Sorting $= \sqrt{\dfrac{Q_3}{Q_1}}$

Skewness $= \dfrac{Q_1 \times Q_3}{(Q_2)^2}$

Cumulative curve

Fig. 51. Grain size curves

Deducing the type of transport from grain size curves alone is a delicate operation. However, some facts are well established. *Dune sands* are in general better sorted than fluviatile sands. *Marine beach sands* are low in fine particles and are weighted towards the coarser grain sizes (negative skewed frequency curve), while fine particles are more important in fluviatile or aeolian environments (positive skewed frequency curve).

Different minerals show differing resistances to abrasion: the easily cleaved felspars are less resistant than quartz.

Folk has defined four stages in the alteration of a sediment after derivation from a weathered source rock (Fig. 52):

— *an immature stage* where the sediment is rich in clays and fine micas; the larger particles are still angular and poorly sorted;
— *a submature stage* where the fine particles have mostly been winnowed out; the grains are still angular and poorly sorted;
— *a mature stage* is reached when clays are entirely eliminated; the particles are well-sorted but still subangular;
— *a supermature stage,* the ultimate in the evolution of the sediment, where the grains are well-rounded and well-sorted; there are no clays.

Sediments reach one or other of these four stages of evolution depending on the degree of hydrodynamic energy to which they have been subjected.

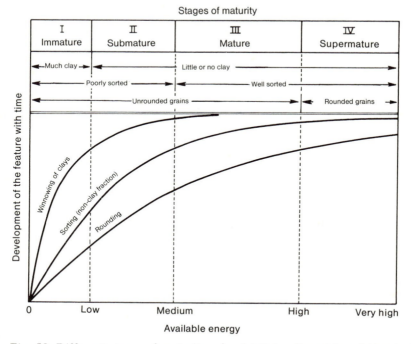

Fig. 52. Different stages of maturity of a detrital sediment in relation to the energy available during transport. (Folk 1951)

b) Carbonate Rocks

Carbonates are transported either as fragments or in solution. In the first case, they behave exactly like detrital particles. In the second case, they are deposited either chemically or through the action of living organisms.

Depending on the size of the calcite crystals, classically one can distinguish:

— **micrite** whose elements are always less than 10 microns in diameter (Fig. 53); it is formed from chemically or mechanically derived carbonate muds and thus indicates a calm, current-free environment. Micrite is the chief constituent of lithographic limestone.

— **sparite** formed by larger (greater than 10 microns) clear crystals precipitated in the spaces between the recongisable debris (Fig. 53). This carbonate cement develops when micrite does not fill the intergranular voids. Its presence implies a high energy environment of deposition. However, sparite can also form by recrystallisation of smaller crystals.

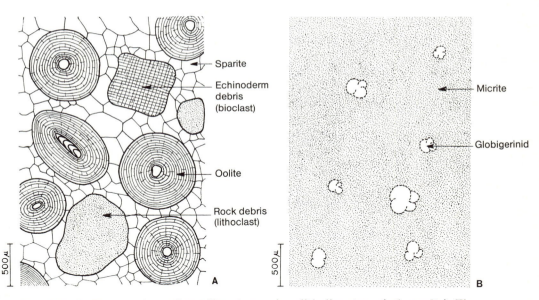

Fig. 53A, B. Thin sections of two limestones. **A** oolitic limestone (calcarenite). The rounding of the grains, the sorting and the presence of sparite indicate a high energy environment. **B** Chalk (calcilutite). The micritic character of the rock and the presence of planktonic foraminifera indicate a quiet water deposit

In general, **detrital sediments** can tell us about their **source area** and their **means of transport**, while **chemically** or **organically formed rocks** tell us more about the **nature** of the **environment of deposition**.

4. Particle Distribution

a) Graded Bedding

As the energy level of the transporting medium progressively drops, particles are deposited in decreasing order of size and graded bedding is formed. When this happens, there is a vertical **decrease in grain size** (Fig. 54). If, on the other hand, energy increases with time, there is a resulting **increase in grain size**. This grading can be vertical or lateral (Fig. 55). Frequently grading is seen by the naked eye as a colour change: the coarse part is lighter than the fine part, which is richer in clay minerals and organic matter.

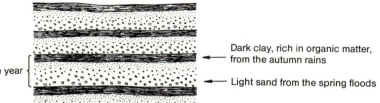

One year

Dark clay, rich in organic matter, from the autumn rains

Light sand from the spring floods

Fig. 54. Varves: fining-up periglacial sediments (each sequence is several millimetres thick)

Graded bedding develops best in water. Glacial deposits, on the other hand, are characterized by poor grading due to the high viscosity of the ice. Clays full of boulders are deposited in bulk during thaws — the geological equivalents are called *tillites*.

b) Parting Lineation

Parting lineation can be seen on bedding planes in grits as elongated splinters with perfectly parallel long axes (Fig. 56). This flaking is the result of the orientation and segregation of particles during deposition. It only occurs in a high energy environment.

c) Pebble Orientation

Elongated pebbles are aligned by the action of currents. On beaches, the long axes of the pebbles are aligned parallel to the shore by wave movement. In a fluviatile environment, the long axes of the pebbles have a

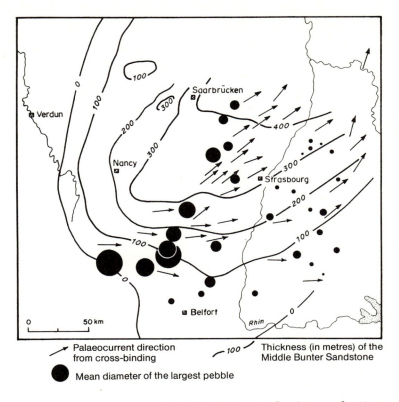

Fig. 55. Reconstruction of palaeocurrents in the Middle Bunter Sandstone of eastern France and southern Germany from the lateral grading of the pebbles and cross-bedding measurements. (Forche 1935)

Fig. 56. Parting lineation on the surface of a slab of sandstone. *Arrow* indicates current direction

tendency to be aligned either parallel to or perpendicular to the direction of flow. Irregularly shaped pebbles are anchored in the sediment by their largest part: the protruding part points downstream.

d) Pebble Imbrication

When flat pebbles are deposited by a unidirectional current, they lie like tiles of a roof with their long axis dipping upstream. This orientation is frequently seen in fluviatile deposits (Fig. 57).

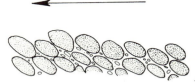

Fig. 57. Pebble imbrication in a fluviatile deposit. *Arrow* indicates current direction

II. Observations on Beds: Stratinomy

1. Stratification and Bedding

Stratification is the arrangement of sedimentary deposits in separate horizons, the beds or strata. They are separated from each other by bedding planes or by interbedded strata of a different lithology. Each bed represents a sedimentary episode.

Within these beds, geometrical structures which are produced during deposition can be seen.

a) Stratification

From the geometry, two types of stratification can be distinguished:
— **planar parallel stratification** where the surfaces which define the beds are noticeably parallel over long distances. The deposits may thus have a lateral extent of the order of kilometres or even tens of kilometres. They form beneath bodies of water which have a great lateral extent: seas, lakes, etc.;
— **lenticular stratification** where each bed is shaped like a lens whose extent ranges from a few decimetres to several tens of metres (Fig. 105). These desposits are thus restricted in space: depression or channel fills, dunes reefs, etc.

Fluviatile or **tidal channels** are defined by a basal erosion surface which is irregular and generally concave (Fig. 76). Usually they are filled in several stages. They often contain locally derived rock fragments.

b) Bedding

Bedding is most often seen as a succession of sedimentary layers on a millimetre scale called **laminae**. These develop as a result of fluctuations in the rate of deposition or in the type of sediment supplied. They can be graded. Care must be taken not to confuse sedimentary bedding, which is of mechanical origin and contemporaneous with deposition, with cleavage, which is tectonically formed, or with the banding caused by algae (stromatolites).

Within the beds, the arrangement of the laminae gives different forms of bedding:

i) Horizontal Bedding

The surfaces of the laminae are planar and parallel to the top and bottom of the beds. In a fine-grained sediment (clay, micrite), horizontal bedding is produced by simple settling-out of particles: it indicates a quiet environment. In contrast, in a coarse-grained sediment (grits), it occurs because of the action of currents, in a high energy environment.

ii) Cross-Bedding

The laminae form an acute angle with major bedding planes (Fig. 58). This type of bedding develops in a unidirectional current system. The transported grains build an oblique slope on which successive laminae are

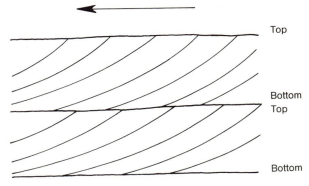

Fig. 58. Cross-bedding in two successive beds. *Arrow* indicates current direction

deposited, dipping downstream. These oblique beds can indicate current
direction. In water, the laminae meet the underlying bed at an angle of
less than 34°, generally nearer 20°. In wind-blown deposits, this angle
may be higher, up to 40°.

Cross-bedding is produced by rhythmically repeated sedimentation:
ripple marks, aeolian or water-laid dunes, fluviatile bars, etc. Depending
on their size, one can get small-scale cross-bedding (ripple marks; unit
size of a few cm) which is often mis-called "**microcross-bedding**" or large-
scale cross-bedding (dunes, bars; unit size ranging from a few decimetres
to metres). In the second case, depending on the shape of the units
(Fig. 59), one can identify:

— **oblique tabular stratification** where the boundaries between units are
  planar surfaces; these result from the movement of an underwater bar;
— **festoon bedding** where the surface between the units is spoon-shaped;
  these result from the movement of major ridges or dunes.

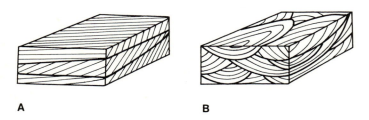

A   **B**

Fig. 59A, B. Types of cross-bedding. **A** tabular cross-bedding; **B** curved cross-bedding

iii) Flaser Bedding

The movement of ripple marks can produce a mottled structure, flaser
bedding (Fig. 60). When an area of ripple marks stops being active, the
fine material which had been in suspension is deposited in the hollows
between ripples. A new development of ripple marks will erode the crests
of the preceding set, but generally not touch the valleys. In section, the
sediment shows small concavo-convex lenses of fine material within a
coarser deposit. This type of deposit is particularly frequent in the intertidal zone and in estuaries where ripple formation alternates with periods
of slack water. The opposing dips of successive horizons indicates current
reversal (herringbone structure).

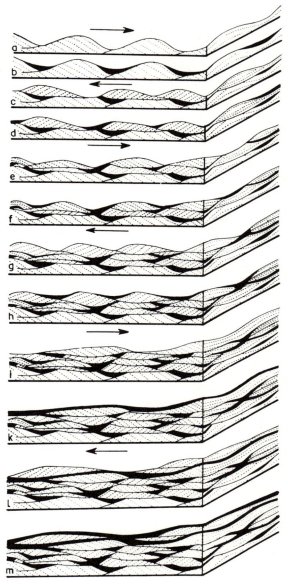

Fig. 60. Diagram showing the origin of flaser bedding from successive trains of ripple marks. *Arrows* indicate the reversal of current direction from flood *(a, e, i)* to ebb *(c, g, l)* separated by periods of slack water *(b, d, f, h, k, m)*. *Black* pelitic deposits; *dotted* cross-bedding within the ripples. (Reineck 1960)

iv) Convolute Bedding

Shortly after being deposited, some horizons slip gently down-slope (a slope of 3° is sufficient). This produces a set of rather irregular convolutions between less plastic horizons which remain planar (Fig. 61). These sedimentary microfolds (convolute bedding) occur in sediment which is saturated with water, in other words, rapidly deposited.

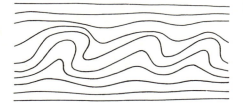

Fig. 61. Convolute bedding within a bed

2. Sedimentary Structures

Sedimentary structures are small-scale geometrical features at the surface of or within a bed, resulting from the action of a fluid (water, wind).

a) Structures on the Top of the Beds

i) Ridges or Ripple Marks

Ripple marks are rhythmically repeated undulations formed on mobile sediment by currents or swell. The wave length (distance between two crests) varies from a few centimetres (ripple marks) to some tens of metres (dunes).

There are two types:

— **oscillation ripples,** with a symmetrical profile (Fig. 62): formed in shallow water by the back-and-forth movement of the swell; they indicate shallow water;

— **current ripples,** with an asymmetric profile (Fig. 63): formed by the transport of sedimentary grains by a unidirectional current which transfers material from the upstream slope to the steeper downstream face of the ripple. Their crests lie at right angles to the direction of current movement and may be linear or discontinuous (Pl. II, Fig. 2). These ripples are not characteristic of any particular water depth, though they are most common in shallow water; the direction and sense of palaeocurrents may be deduced from them.

Theoretically, wind-formed ripples can be distinguished from water-formed ripples since, under water, the coarsest grains tend to accumulate in the hollows while in the wind they congregate near the crests.

Successive fields of ripple marks give rise to **flaser bedding,** which is common in shallow water (Fig. 60), and to **microcross-bedding** (Fig. 64).

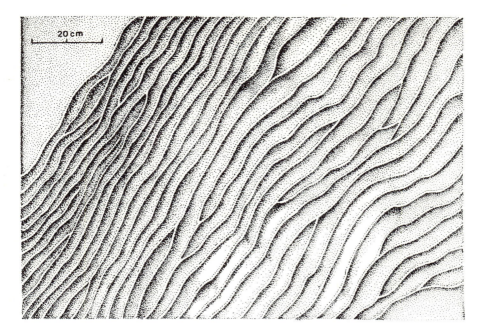

Fig. 62. Oscillation ripples on the surface of a sandstone horizon

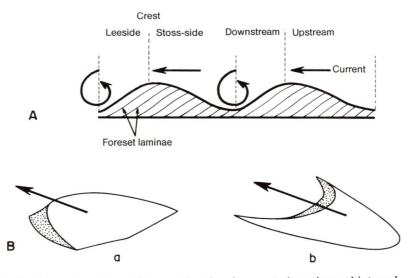

Fig. 63A, B. Ripple marks produced by a unidirectional current. **A** section and internal structure; **B** simple ripples with discontinuous crests. *a* linguoid ripple; *b* crescent-shaped ripple. (Gall 1971)

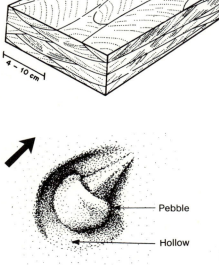

Fig. 64A, B. Block diagram showing the origin of micro-cross-stratification by ripple growth. A upper surface of the ripples (*arrow* indicates current direction). B subhorizontal section. (Wurster 1963)

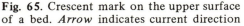

Fig. 65. Crescent mark on the upper surface of a bed. *Arrow* indicates current direction

ii) Crescent Marks

Crescent marks are horseshoe-shaped depresssions which form upstream of a small obstacle (pebble, shell) by current scour (Fig. 65). The convex end points upstream.

b) Structures on the Base of the Bed

On the base of beds, sedimentary structures (sole marks) occur as casts, either of depressions or in relief, of structures on the upper surface of a fine-grained underlying bed. They are classified according to their origin:

i) Structures Indicating the Physico-Chemical Environment of the Underlying Bed

- **desiccation cracks** have a characteristically polygonal shape; they indicate drying-out of a wet environment;
- **salt pseudomorphs** are the infill of holes left after the solution of halite crystals (Fig. 66); they indicate a highly saline environment.

ii) Structures Made by Currents

- **scour marks** are hollows eroded in a mobile sediment by the current. Among the commonest are **flute marks**, which have **flute casts** (Fig. 67) as their counterpart on the base of the overlying bed. The rounded, bulb-shaped end points upstream and indicates the direction and sense of the palaeocurrent. They generally occur in groups;
- **rill marks** can develop in shallow water, as a finely ramified grouping of little furrows;
- **tool marks** are made by objects carried along by the current and vary according to the type and angle of impact of the object (Fig. 68). **Groove marks** (Figs. 69, 93) result from something being dragged along the surface of the sediment; plants, dead animals, pebbles. They indicate current direction but not usually its sense.

66

2 cm

67

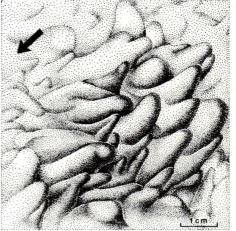

1 cm

Fig. 67. Flute casts on a bedding plane. *Arrow* indicates the direction of the palaeocurrents

Fig. 66. Salt pseudomorphs on a bedding plane

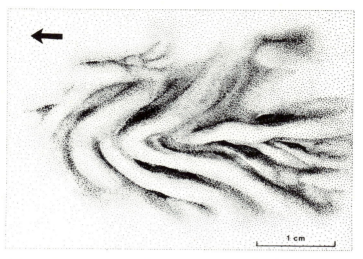

Fig. 68. Prod mark made by an object carried by the current

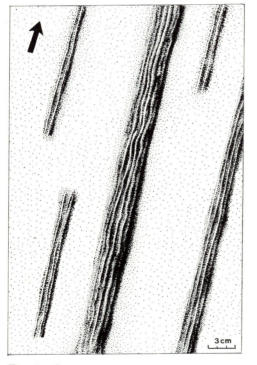

Fig. 69. Groove casts on a bedding plane

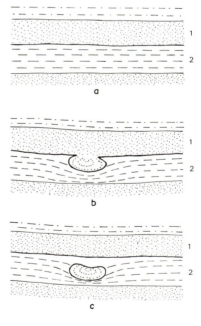

Fig. 70. Origin of a pseudonodule within a hydroplastic sediment. *1* sand; *2* clay

iii) Deformation Structures

During compaction, sediment can be intruded into a highly water-saturated underlying bed, forming **load casts** and **pseudo-nodules** (Fig. 70).

c) Structures Within Beds

Within beds, one sees many structures formed by the orientation of grains (**parting lineation**; Fig. 56) or by the deformation of the sedimentary laminae (**synsedimentary microfolds** if the horizontal component of movement is greater than the vertical; **pseudonodules** if the vertical is greater than the horizontal). Other structures appear later, after compaction of the sediment, by diagenesis. Their relationship to the original environment is often difficult to prove. **Concretions** of various types (geodes, septaria, ovoids, etc.) can be formed. **Stylolites**, zig-zag structures which are common in limestones (Fig. 71), are caused by the solution of carbonates during compaction or tectonism.

5mm

Fig. 71. Stylolites in a limestone

3. Conclusions

Sedimentary structures are a fruitful area for learning about the deposition of sediments and the physico-chemical conditions in sedimentary environments.

By means of experiments, both in the field and in the laboratory, one can understand the relationship between **hydrodynamic energy**, the transport of grains and the sedimentological character of the deposit (Figs. 72, 106). Thus, for a given grain size, a current will successively cause ripple marks, underwater dunes and then planar laminae with parting lineation. In highly turbulent suspension currents, **antidunes** with laminae dipping upstream are formed, but are rarely preserved; in these conditions, erosive features and sole markings often occur.

Sedimentary structures	Bedding structures		Sediment transport	Energy level
Frequent bioturbation	Planar horizontal laminae	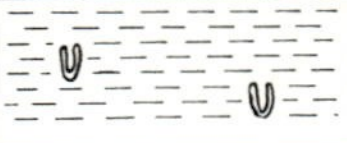	Settling-out of fine particles	
Short wavelength ripple marks	Small scale cross-bedding dipping downstream	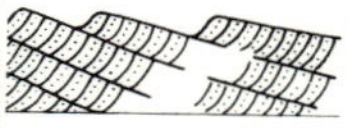	Transport by traction	Low
Long wave-length current ripples (underwater dunes)	Large scale cross-bedding dipping downstream	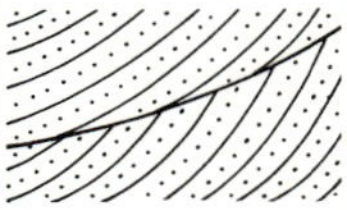	Transport by traction	
Parting lineation	Planar horizontal laminae	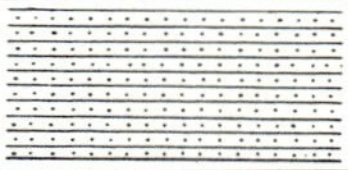	Transport by traction	
Anti-dunes (rarely preserved)	Cross-bedding dipping upstream	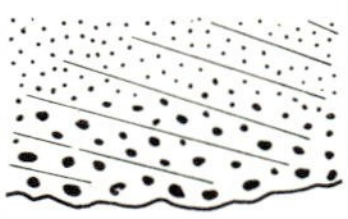		High
Erosion surface with tool marks and flute marks	Vertical grading		Transport in suspension	

Fig. 72. Sequence of sedimentary structures induced by a unidirectional current of decreasing competence *(from bottom to top)*. This table is valid for a given grain size and a constant depth of water

III. Observations at Outcrop: Analysis of Sequences

At outcrop, it is often possible to discover some logic within the apparent anarchy of the vertical succession or horizontal distribution of beds. In a geological succession, the establishment of **sequences** or the different units of which they are made up, without major discontinuity, in a precise, rhythmically repeated order, can help one to understand the processes of sedimentation.

1. The Idealised Normal Sequence

The idealised normal sequence, defined by Goldschmidt and modified by Millot, lists the succession of deposits normally found in a sedimentary basin close to an area of high relief undergoing erosion. It consists of:

a) coarse **insoluble residues**, pebbles and sands, produced by intense mechanical erosion of high ground;

b) **colloidal hydrated minerals (hydrolysats)** — the fine clay fraction, removed by rivers flowing slowly over low relief.

These two units of **detrital sediments** are transported by the water as actual objects.

c) **oxides**, especially sedimentary iron and manganese, formed as the high ground is peneplaned;

d) **carbonates** of calcium and magnesium which produce limestones and dolomites;

e) **salt deposits** formed in unreplenished bodies of water; they represent the final stages of planation of the land.

These three units are **chemical sediments** since they are transported in solution.

2. Lithological Sequences

The five-unit ideal sequence is hardly ever seen in nature. Only small parts of it are seen in geological outcrops: these are **lithological sequences.**

a) Positive and Negative Sequences

Depending on the order in which the units of a lithological sequence appear at outcrop, Lombard identifies two types of sedimentary evolution (Fig. 73):

— **positive sequences** which follow the order of the ideal succession: vertically, one passes from detrital sediments to chemical sediments. This is a characteristically **transgressive** sequence;

— **negative sequences** which reverse the order of the ideal succession: they characterise **regressions.**

A classic example of a positive sequence linked to a transgression is that of the Germanotype Trias: highly simplified, it passes from the red sandstones of the Bunter, through the marls and limestones of the Muschelkalk to the gypsum and rock salt of the Keuper.

b) Sequences Derived from Soils

Erhart's theory of **biorhexistasy** emphasises the role of the forest in protecting soil from erosion and controlling the removal of the products of terrestrial alteration.

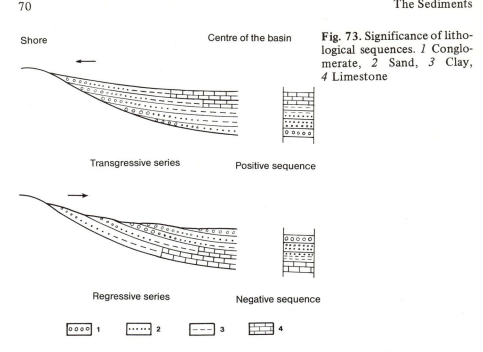

Shore   Centre of the basin

Fig. 73. Significance of lithological sequences. *1* Conglomerate, *2* Sand, *3* Clay,
4 Limestone

Transgressive series   Positive sequence

Regressive series   Negative sequence

In a period of **biostasis**, the forest acts as a filter: insoluble particles
are not removed and only soluble products reach the sedimentary basin.

In a period of **rhexistasis**, the destruction of the forest frees detrital
sediments to be washed downstream.

The succession of these two phases produces a negative sequence in
the sedimentary basin as one passes from chemical to detrital input.
Desprairies has described similar sequences from the lacustrine Stampian
of southern Limagne (Fig. 74).

3. Rhythmic Series

Some sediments are made up of sequences which are repeated many
times (rhythmic or cyclic series); this can occur on various scales. It is
always caused by the recurrence of the factors controlling sedimentation.

a) Varves

Varves are the sediments of periglacial lakes and seas. They are formed
by millimetre-scale repetitions of two units:
— light-coloured sandy sediments from the spring floods;
— darker, organic-rich clays from the autumn rains.

Fig. 74. Sequence of pedological origin in the Eocene and Oligocene of south Limagne. *1* limestones; *2* red kaolinitic clays; *3* sands. (Desprairies in Aubouin 1968)

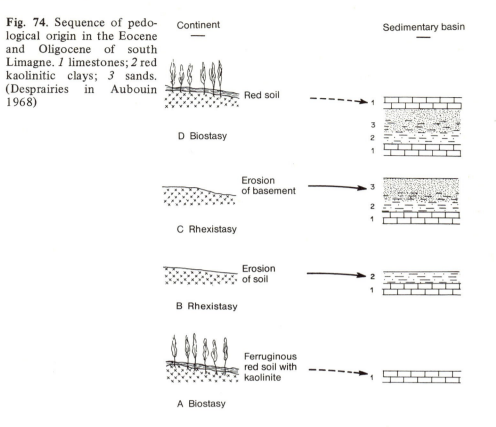

Each sequence represents a year. Thus De Geer was able to establish an absolute chronology for post-glacial times in Scandinavia. Varved sediments can be found in lakes outside the glacial region; these are also associated with seasonal variation.

b) Alternation of Limestones and Marls

Alternating sequences of limestone and marl are common in geological successions, especially in the shallow seas of the Triassic and Jurassic, and can be explained by a variety of processes.

i) Hallam's Theory of Eustatic Control

Hallam interprets limestone-marl alternations as due to periodic variations in sea level. Chemical or biochemical precipitation of calcium carbonate is at a maximum in shallow seas where aeration, temperature and light conditions are optimal; deposition of clays is favoured by deeper water where movement is less.

When sea level drops, limestones with a rich fauna are developed. Their tops show signs of emergence. When sea level rises again, faunally impoverished marls and clays are deposited. Sometimes the water even becomes stagnant and an anaerobic environment develops (bituminous shales).

Hallam's ideas have been satisfactorily applied to the Lias of Western Europe.

ii) Lombard's Theory of Gravity Flow

According to Lombard, carbonates accumulate in shallow water. Occasionally, because of the sediment's own weight, equilibrium is disturbed and the sediment begins to move downslope, finishing up in great sheets in the bottom of sedimentary basins. Several types of sedimentary structure provide evidence for this; the fossils also show signs of mechanical abrasion. Thus the carbonate horizons and their faunas are allochthonous. Meanwhile, clays continue to be deposited, forming a **"background noise"** to the sedimentation, and giving rise to marly horizons with an impoverished, autochthonous fauna.

The best example of this are the calcareous deposits of the Alpine geosyncline.

In the Upper Muschelkalk of eastern France, Haguenauer has described positive (i.e., fining-up) sequences some decimetres in thickness, separated by erosion levels (Fig. 75). The lower part consists of limestone rich in

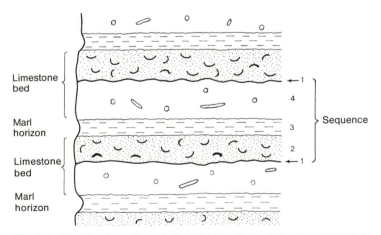

Fig. 75. Rhythmic sedimentation in the Upper Muschelkalk of Alsace and Lorraine (thickness of the sequences: several decimetres). (Modified from Haguenauer 1963) *1* Erosion surface chemically precipitated argillaceous limestone with an autochthonous fauna (burrows); *2* Calcarenite with allochthonous fossil debris (shells, crinoid ossicles); *3* Erosion surface; *4* Sequence

organic debris (crinoid ossicles, broken shells, etc.), often graded and cemented by micrite: a biocalcarenite. This deposit is allochthonous. The middle part of the sequence can be marly, resulting from the deposition of the fine fraction of the detrital material. The top of the sequence is a micritic argillaceous limestone with some burrows, indicating an indigenous fauna. In this particular example, the bedding planes do not necessarily correspond to the units of the sequence. The deposits are formed in shallow water, in a lime-rich environment where precipitation of carbonates as micrite is continuous. Occasionally, coarser sediment is carried into the environment of deposition from neighbouring areas where there is a rich fauna, which disturbs the continuous "background noise" of clay and micrite deposition.

iii) Climatic Variations

Seibold has shown that, in the Upper Jurassic of Germany, the alternation of carbonates and marls occurred in an environment where argillaceous sedimentation was continuous, but periodically interrupted by a massive precipitation of calcium carbonate, which can be attributed to phases of climatic amelioration which favour carbonate precipitation.

iv) Diagenetic Origin

In some cases, the alternation of carbonates and marls can be attributed to a late stage mobilisation of carbonates in the original mud, resulting in horizons rich in limestone, which is often **nodular**.

c) Molasse

Molasse is a mainly detrital sediment resulting from the erosion of a recently emerged landmass. It is typical of the late stage of an orogeny.
   The deposits are often immature grits with a carbonate cement. Beds are generally lenticular, deposited in a transitory sedimentary environment: fluviatile channels, lakes, shallow embayments of the sea, etc. Bersier interprets the rhythmic nature of the Aquitainian and Chattian molasse of Switzerland as being caused by **meandering** of streams in an alluvial plain, producing a successive build-up of sediment. The progressive decrease in carrying power of the current because of infilling of the channel produces **cyclothems**, which ideally comprise (from bottom to top): graded grits, marls, limestones and coal (Fig. 76). The upper parts

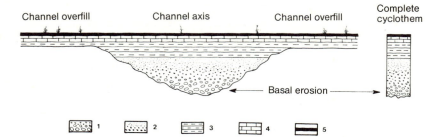

Fig. 76. Origin of fluviatile sequences in molasse. (Modified from Bersier 1958). *1* Calcareous sandstone *(molasse); 2* Argillaceous calcareous sandstone *(macigno); 3* Marls and clays; *4* Lacustrine limestone; *5* Coal near Lausanne. (Bersier 1958)

Stratigraphic section of a cyclothemic series

Dissection of the series into successive cyclothems

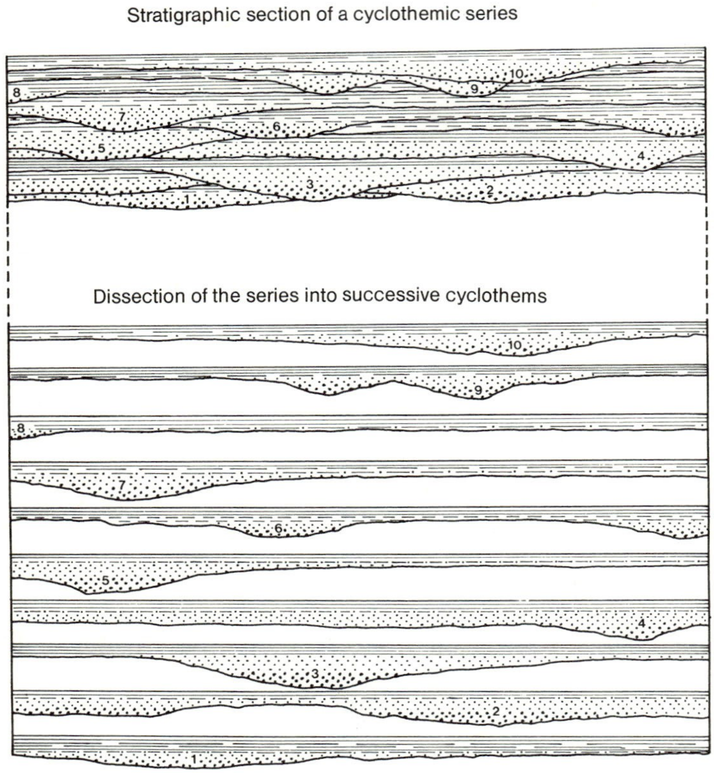

Fig. 77. Interpretation of fluviatile sequences (cyclothems) in the Aquitainian molasse near Lausanne. (Bersier 1958)

of a cyclothem can be cut off by erosion. Alternatively, they can be stunted, with the coarse parts at the base missing, as far up as the channel overfill (Fig. 77). As rivers continually carry detrital material over their own deposits, so they cut into their own banks to make a new channel, and thus spread a more-or-less even load of sediment over a subsiding alluvial plain.

The red sandstones of the Permo-Triassic are comparable to the Tertiary molasse of the Alps. The Bunter Sandstone of the Vosges has two main constituents: a coarse channel deposit and clays from temporary pools (Fig. 81).

d) Flysch

Flysch is a pre-orogenic or syn-orogenic sediment, made up of the fill of subsiding sedimentary troughs preceding the rise of a mountain chain.

The exact composition depends on the nature of the terrain undergoing erosion: there is sandy flysch, calcareous flysch, argillaceous flysch, etc. These sediments are always low in organic material and are usually complex and of wide horizontal distribution. Characteristically, they are graded and show a wide variety of sedimentary structures (Fig. 85). They are very thick and usually laid down in very deep water. It is generally accepted that flysch is deposited by **turbidity currents**: high density flows of water which come off the continental shelf and drop their sedimentary load in deep basins where they are graded. Generally they originate in regions of tectonic instability. The lower part of these flows erodes the underlying sediment and brings in a derived fauna. The upper part, on the other hand, represents the "background noise" sedimentation and contains autochthonous microfossils and many trace fossils.

Although the word "flysch" was originally limited to Alpine deposits, nowadays it is considered as a phase in the evolution of a geosyncline. The word **Culm** is its equivalent in the European Hercynides.

e) Coal

Coal forms in subsiding lacustrine or marine basins, usually on the edge of an orogenic chain. The subaerial accumulation of plant material results in the formation of coal seams, while the associated sands and clays are deposited in shallow water.

There are two basic units, which have a limited horizontal extent: the **sterile** beds (sands, clays, etc.) and the **coal seam** (Figs. 78, 105). In the Westphalian of the North of France, there are 400 cycles within a sedimentary thickness of 2000 m.

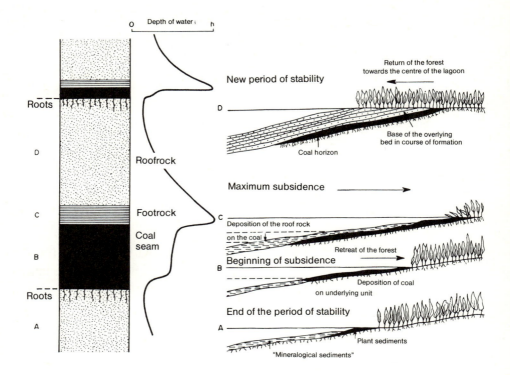

Fig. 78. Origin of cycles in the Carboniferous of north France. (Pruvost in Aubouin et al. 1968)

Many mechanisms have been invoked to explain the cyclicity of coal sequences: rapid irregular subsidence periodically destroying the plant cover; river diversion suddenly flooding an alluvial plain, alternation of dry and wet climates, etc.

To summarize, rhythmic sequences can be explained by various means:
- diversion of fluviatile channels over a broad, subsiding area; for example: molasse, coal;
- by tectonic jerks or subsidence which varied in speed; for example: some coal;
- by alternation in sea level produced by deformation of the ocean bottom or possibly by the alternation of glacial and interglacial phases; for example: limestone-marl alternations;
- by climatic pulsations or the alternation of seasons; for example: varves, some evaporites;
- by gravity flow on underwater slopes; for example: flysch, some limestone-marl alternations;
- by diagenetic modification; for example: some limestone-marl alternations.

All these observations on grains, beds and outcrops must be supplemented by **regional studies** by which we can follow geographical variation and correlation of facies. This detailed analysis will eventually enable us reconstruct the appearance and activity of ancient environments.

Chapter 5   Sedimentary Environments

In the reconstruction of ancient sedimentary environments, one uses information gathered both from fossils and from the sediment itself. From this one can visualise the infinite variety of environments which have occurred during the long history of the Earth. One can group these environments into a limited number of categories.

I. Continental Environments

Continental environments are largely controlled by two climatic factors, temperature and humidity, and may be distinguished from marine environments by several characteristics:
— the presence of terrestrial or freshwater fossils;
— abundant emersion structures such as desiccation cracks;
— the occurrence of fossil soils or coal deposits;
— frequent red colouration (iron oxides) resulting from oxidising conditions on land;
— indications of wind or glacial erosion;
— a predominance of detrital over chemical sediments;
— low or fluctuating palaeosalinities.

1. Fluviatile Environments

a) Alluvial Cones

When streams and rivers flow into a valley or onto a plain, they drop their load as alluvial cones or flood plains. These generally build out from the base of the uplands which provide the source of the sediment and have various characteristics:
— there is little alteration of the sediment because of its violent and rapid erosion and transport;
— the sediment is coarse and poorly rounded because of its local origin and short transport: hence it is **immature**;
— sorting and bedding are poor since the material is simply dumped.

b) Fluviatile Deposits

The type of sediment carried by streams and rivers varies greatly depending on the climate and topography. However, fluviatile deposits do have
the following things in common:
- **bedding is lenticular** since the sediments are deposited in channels or
  pools of water; curved cross-bedding is common;
- **flow directions** indicated by sedimentary structures are **unidirectional**
  and generally constant (Fig. 55);
- sedimentary sequences typically show a decrease in the carrying power
  of the current with time: the base is erosional, there is a clear upwards
  grading, horizontal bedding is succeeded by cross-bedding, the size of
  sedimentary structures decreases towards the top of the sequence, etc.
  (Fig. 106).

Two main facies can be seen in fluviatile sequences (Figs. 79, 80, 101):
- the **channel facies** where the coarsest sediments are deposited at times
  of flood; organic remains are rare and broken;
- the **sheet facies**, deposited in pools of quiet water formed during
  periods of low flow or in areas sheltered from strong currents; the sediment is fine-grained and the area was populated by plants and animals.

The river system can vary between two extremes (Fig. 81):
- a **braided river system**: the slope is high and the sedimentary load
  heavy. The courses of the channels change frequently during successive
  sudden floods, but the spread of palaeocurrent directions is low. The
  deposits are generally coarse;

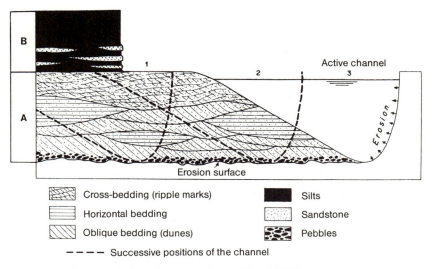

Fig. 79. Deposits of a meandering river complex. (Allen 1971)

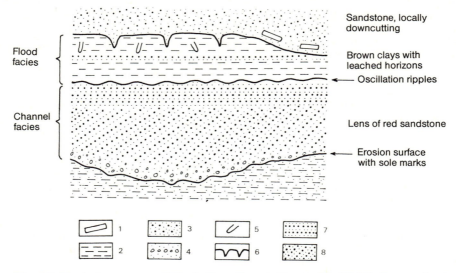

Fig. 80. Typical sequence of fluviatile sedimentation in the Middle Bunter Sandstone of the Vosges (the succession is several metres thick). *1* clay pellet, *2* clay, *3* sandstone, *4* quartz- and quartzite pebbles, *5* Lebensspur, *6* desiccation cracks, *7* horizontal bedding, *8* cross-bedding

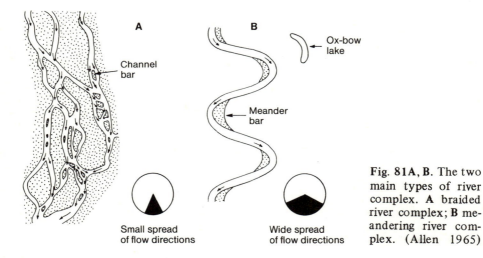

Fig. 81A, B. The two main types of river complex. **A** braided river complex; **B** meandering river complex. (Allen 1965)

— a **meandering river system**: water flow is gentle and carries a fine-grained load. Deposits collect on the convex banks of channels and form **point bars** which clearly show a decrease in energy level with time (Fig. 79). Fossil soils or coals are frequent, and the palaeocurrent directions are more variable than in a braided channel system.

For any given river, the braided regime can be succeeded by the meandering regime both from upstream to downstream and in the course of time.

2. Lacustrine Environments

Lacustrine sedimentation depends on the climate and the amount of water entering the system. Deltas apart, bedding is planar parallel. The deposits tend to form **concentric rings** of decreasing grain size from the banks to the centre of the lake. Quiet water favours clay or limestone deposition.

Climatic variations cause major variations in the geochemical composition of lake waters and seasonal variations can cause varves. A period of dry climate causes stagnation of the water with the precipitation of evaporites or an increased concentration of some minerals, mainly iron.

Although lacustrine and fluviatile sediments are often associated, it is not difficult to distinguish between them.

The **lacustrine fauna** is low in diversity compared to a marine fauna. It consists chiefly of freshwater bivalves and gastropods together with fish. Burrowers are generally rare because the sediment is poorly oxygenated. **Plants** are especially abundant close to the banks.

3. Aeolian Sedimentation

Wind-blown sediments, chiefly dunes, are difficult to distinguish since they consist of mobile sediments deposited by air, easily reworked by water.

They are mostly sands, with a grainsize varying between 0.06 and 2 mm, the clay grade fraction and the micas having been winnowed out. Wind transport produces characteristically shaped particles: the grains are round and matt, while pebbles are faceted (Windkanters; Fig. 49). Grading is always good. Bedding consists of thick groups of crossbedded units, often concave upwards, whose dip may be as high as 40°. Dip directions vary greatly, however, because of variations in wind direction. Ripple marks form on the dune surfaces, but are never oscillation-type.

During the Quaternary, **loess** was formed from the deposition of fine dust with a grain size between 0.015 and 0.045 mm, often carried by the wind for great distances. It is typically unbedded, though it may contain the remains of pulmonate gastropods and vertebrate bones.

4. The Glacial Environment

A **glacial geomorphology** is very distinctive: U-shaped valleys, striated surfaces, roches moutonnées, etc. The sediments are of local origin. Because of the low temperatures, delicate minerals such as felspars and the ferromagnesians have not been altered. The grains are angular and the sediment is **immature**. Sedimentary material transported as moraines, often reworked by glacial streams, is characterised by a total absence of grading: boulders and pebbles are often angular, striated and lie within a fine-grained matrix (Fig. 50). These boulder clays are the origin of **tillites.**

Glacial periods can equally well be recognised by periglacial phenomena: soil polygons, cryoturbation structures, ice wedges, etc.

II. Marine Environments

Because of the variations in depth and turbulence of different bodies of water, marine environments are very variable (Fig. 82), but they all have several common characteristics which distinguish them from continental environments:

— a diverse marine flora and fauna;
— the absence of emersion structures (except in the intertidal zone);
— a constant palaeosalinity (except in lagoons);
— chemical and biochemical sediments (especially limestones) predominate over detrital sediments;
— a horizontal continuity of beds formed beneath permanent and extensive bodies of water;
— a generally low dip of sedimentary units and the development of tabular rather than concave-up cross bedding;
— an almost complete absence of red colouration due to iron oxides.

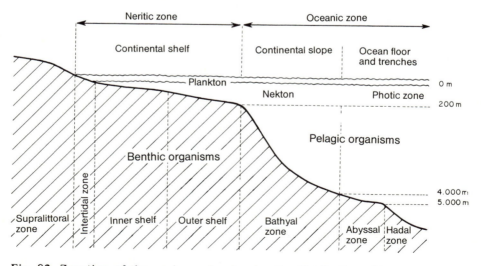

Fig. 82. Zonation of the marine realm showing the distribution of living organisms

1. The Intertidal Zone or Shoreface

The marine marginal areas which are subject to the tides, including tidal flats, form the intertidal zone. It is a high energy area where the sedimentation rate is high and the sediments are continually reworked. Evidence of emergence is common.

a) The Sediments

The sediments consist of sands and clays which are thinly but often irregularly or obliquely bedded. **Current ripple marks** are the commonest sedimentary structure and create bedding features such as flaser bedding or laminar cross bedding, reflecting the oscillating currents (Fig. 60). Bubbles of air caught in the sediment during periods of emergence could cause the characteristic **birdseye** structure.

In the **supralittoral zone**, which is only rarely flooded, desiccation cracks are formed. In an arid climate, dolomitisation and the precipitation of evaporites can occur, and, if these are later dissolved, an intraformational breccia may be formed.

Intertidal sediments show a horizontal zonation: they are muddier in the higher parts and sandier in the lower parts of the shoreface. They are often cut by **tidal channels** which have a coarser, often cross-bedded, infill, indicating the alternating directions of the tidal currents.

Periodic changes in sea level result in sequences grading between sands which represent periods of maximum transgression and marsh deposits which indicate a long period of emergence.

b) The Organisms

The fauna includes a large number of burrowing animals which protect themselves against desiccation by burying themselves in the sediment where they produce an intense **bioturbation**. They react to a high rate of sedimentation by moving rapidly upwards, which results in a large number of vertical burrows.

The **faecal pellets** produced by these animals form a considerable proportion of the sediment. The surface of the sediment can be carpeted by algae, whose fossil equivalents were responsible for building **stromatolites**. In a tropical climate, the intertidal zone can be colonised by the higher plants, forming **mangrove** swamps.

2. Neritic Environments

The neritic zone extends across the continental shelf and over the whole region is rarely more than 200 m deep. It coincides with the **photic zone** which is penetrated by sunlight and where active photosynthesis takes place. In this region, the fauna reaches its maximum density and diversity.

Most marine sediments accessible to the geologist are shallow water deposits which have encroached on the continental margin by means of transgressions **(epicontinental seas)**. There is a great variety of neritic environments, closely controlled by the type of sediment available and the degree of water turbulence.

a) Neritic Environments Subject to Terrigenous Sedimentation

Detrital material from the continents is reworked by currents and waves. Below a certain depth level, **wave base**, these hardly disturb the sediment; usually this depth is less than 30 m. Close to the shore, waves break on the bottom and create a zone of very high sediment movement. Sediments here tend to be coarse-grained, and are often distributed parallel to the coast as **bars**, **sand waves** or **barriers**. Both shorewards and offshore of these bars, the sediments tend to become finer-grained.

i) The Sediments

Littoral bars are made up of sand and broken shell debris. In this high energy area, fines are winnowed out and the sediments are wellgraded; they have a high degree of rounding. Offshore, they grade to fine sands and clays.

Within the littoral bars, the sediments are usually cross bedded, with the laminae dipping gently offshore, although larger units are tabular. As sea level changes, the bars migrate over the finer onshore or offshore sediments, producing a vertical increase in grainsize. Thus, in boreholes, one can easily distinguish offshore bars from alluvial deposits, which usually decrease in grainsize upwards. The offshore muds are generally thinbedded, but this is often obscured by bioturbation or even occasionally by major storms.

These offshore muds and littoral sand bars can pass shorewards into lagoonal or intertidal deposits and eventually to continental deposits.

ii) The Organisms

Littoral bars are rich in benthic animals, but their shells are generally broken. The offshore muds contain an abundant endofauna.

b) Neritic Environments with Carbonate Deposition

When sedimentary input from the land is low, carbonates tend to be deposited on the continental shelf. There are three main, depth-controlled environments (Fig. 83):

- **offshore**, finely bedded, calcareous clay (micrite) are deposited in the low energy environment, undisturbed by wave action. The fauna is mainly pelagic. **Glauconite** can form in this environment. If the water is poorly aerated, organic matter may be preserved;
- **above wave base** is a high energy environment where fine-grained particles are winnowed out. This produces a well-graded, coarse-grained sediment with abundant skeletal debris (calcarenite) which accumulates to form a bar, usually cross bedded. In this environment, **onkoliths** and **reefs** can form. Thick-shelled benthic foraminifera are abundant;
- **between the bar and the shore** the water is quiet and shallow, forming a **lagoon** with an impoverished fauna. The sediments are fine-grained: calcareous muds rich in faecal pellets — often chemically precipitated micrite— and even dolomites and evaporites.

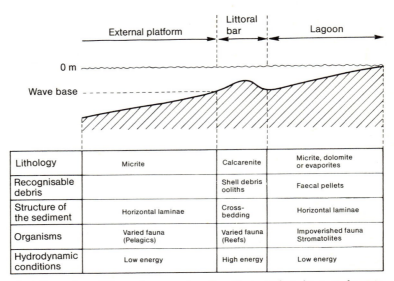

	External platform	Littoral bar	Lagoon
Lithology	Micrite	Calcarenite	Micrite, dolomite or evaporites
Recognisable debris		Shell debris ooliths	Faecal pellets
Structure of the sediment	Horizontal laminae	Cross-bedding	Horizontal laminae
Organisms	Varied fauna (Pelagics)	Varied fauna (Reefs)	Impoverished fauna Stromatolites
Hydrodynamic conditions	Low energy	High energy	Low energy

Fig. 83. Characteristics of neritic environments of sedimentation in a carbonate regime. (Modified from Irwin 1965)

c) Reef Environments

Reefs are calcareous structures built up by the skeletons of fixed organisms growing in situ, and are resistant to mechanical erosion.

i) The Sediments

Depending on their shape, reefs are:
— **bioherms**: large domes surrounded by rocks of a different lithology;
— **biostromes**: more-or-less evenly bedded units.

ii) The Organisms

The reef environment is the site of intense biological activity. Through geological time, various groups of **constructing animals and plants** have formed the building blocks of reefs: encrusting algae (blue-green algal biostromes or stromatolites which began in the Precambrian), the Lower and Middle Cambrian archaeocyathids, sponges, coelenterates (stromatoporoids, Palaeozoic tabulate and rugose corals, post-Permian hexacorals), bryozoans, serpulids, bivalves (Cretaceous rudist biostromes), etc.

As well as the reef builders, there is a **subreefal fauna** living fixed or free within the reef. Their skeletons or shells are thick because of the high availability of calcium carbonate (calcareous algae, brachiopods, bivalves, echinoderms, etc.).

iii) Environmental Parameters of the Reef

Present day reefs, which are built by hexacorals, form in very closely defined conditions: water temperature must be above 18°C, salinity must be normal marine and constant, they must lie within the euphotic zone (less than 40 m), the water must be clear and in constant motion. These requirements are largely imposed by the presence of symbiotic brown algae (**zooxanthellae**) in the soft parts of the corals, who benefit from their photosynthesis.

It is possible that, at other times in geological history, conditions might have varied a little depending on the dominant zoological groups. Thus stromatolites are normally associated with shallow water of variable salinity, while archaeocyathids probably lived in fresher water than present-day corals.

iv) Zonation of Reefs

Present-day reefs may be ring-shaped atolls, or fringing or barrier reefs formed on the continental shelf. These latter types are more common in the geological record. They show a series of zones from the ocean to the coast:
- **fore-reef zone** where reef talus accumulates as a breccia with a micrite cement; cross bedding and debris slips are common;
- **reef core** formed by the build-up of constructing organisms in their life position; erosion by waves and predators produces calcarenite with a micrite or sparite cement. Within the bioherm, one can sometimes see a depth-controlled vertical zonation of organisms;
- **back-reef zone** between the reef core and the shore; sheltered from the waves, a lagoon may develop where micrite, and even dolomite and evaporites, can accumulate.

Within this broad outline, a large number of variations is possible, depending on local palaeogeographic conditions (Fig. 114).

3. The Ocean

The oceanic zone lies far from shore, over the continental slope and the abyssal plains. Sedimentation takes place in great depths of water.

a) The Sediments

Deep ocean sediments are deposited by two very different mechanisms:
- Coastally derived detrital material can be carried across the continental slope to the abyssal plains by **turbidity currents**. This material is **allochthonous**.
- Far from the influence of land and currents, the **settling-out** of fine-grained sediment and the tests of micro-organisms deposits **pelagic muds**, which can be calcareous, argillaceous or siliceous. They give rise to fine-grained, nodular limestones, which are often red (Triassic and Jurassic "**ammonitico rosso**" facies), clays and radiolarites. The latter are often associated with submarine volcanic eruptions.

Pelagic sediments contain abundant **manganese nodules**.

b) The Organisms

Sediments transported by turbidity currents contain an allochthonous fauna transported from shallow water. It is often of a different age from its matrix.

Pelagic muds are largely made up of the tests of planktonic organisms: flagellar calcareous shells (coccolithophorids) and foraminifera (globigerinids), calcareous mollusc shells (pteropods), siliceous tests of diatoms and radiolaria. The macrofauna consists of carnivorous nekton (cephalopods). Benthos is rare and a flora absent because the abyssal depths are well below the photic zone.

Fossil sediments which were undoubtedly deposited in very deep water are extremely rare, and can be easily confused with sediments formed a long way from the shore and sheltered from terrigenous influences: geography can have a similar effect on sediments to depth. Thus the Cretaceous **chalk**, largely made up of the shells of coccolithophorids, used to be grouped with deep-water globigerinid oozes. However, its associated macrofauna (bryozoans, bivalves, sea urchins, etc.) shows that is is really a neritic deposit.

III. Deltas and Estuaries

As they enter a larger body of water, such as the sea or a lake, streams and rivers are slowed down and dump their sedimentary load. When waves and currents are not able to carry away this sediment, a delta is formed and gradually builds out into the sea or lake. On the other hand, when tidal activity dominates the lower reaches of a river, an estuary is formed.

1. Deltas

Deltas are regions of active sedimentation. Their shape depends partly on the quantity of sedimentary input, and partly on the ability of the marine or lacustrine waves to redistribute the sediment.

Upstream, the deltaic environment passes to an alluvial plain (Figs. 84, 111). **Channels** form meanders and abandon ox-bow lakes when the slope decreases. Further downstream, the channels build up **natural levees** from the material they are transporting. These are often breached and a new route established. This builds up the **delta platform**, which is partly submerged. Further on, beyond the break in slope of the **delta front**, the sediments grade to the fine-grained deposits of the sea or lake in the **prodelta** region.

The active part of the delta moves with time and ox-bow lakes, ponds, marshes and lagoons are formed. The delta build-up is eroded by waves and currents and the mouth can evolve into an estuary.

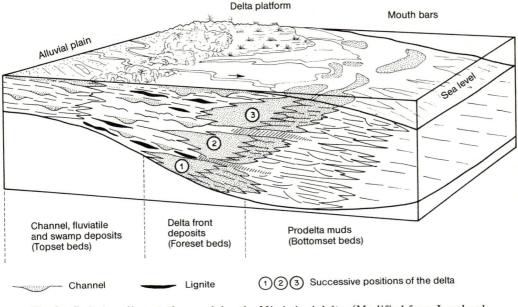

Fig. 84. Deltaic sedimentation model — the Mississippi delta. (Modified from Lombard 1972)

a) The Sediments

— The coarsest sediments are found in the proximal part of the delta, largely as channel fills, looking very like fluviatile deposits. Flanking them are the finer-grained deposits of the levées. The original bedding is often disrupted by plants. Concentrations of iron oxide are common. Clays are often deposited in the lower parts of the flood plain: ponds, ox-bow lakes, swamps, etc.

These sediments formed on the emergent part of the delta are grouped as the **topset beds**.

— In the submerged, distal part of the delta, a great thickness of sediments accumulates, so that the **delta front** builds out towards the sea. This varied group of deposits forms the **foreset beds**. **Mouth bars** form at the exits of the channels, generally made up of interbedded coarse and fine sands rich in plant debris. Clay is partly removed by the waves. Concave cross-bedding is common. Between the bars, finer sediments are deposited.

The speed of sedimentation means that a lot of organic matter is included. As this decomposes, gas bubbles form and rise to the surface, disrupting the bedding laminae. Bedding is often contorted by submarine slumping.

— The **prodelta** clays are rich in organic matter and their fine horizontal bedding is often disturbed by bioturbation. These are the **bottomset beds**.

The deltaic environment is the site of a particularly high **rate of sedimentation**: several tens of centimetres of sediment can be deposited in a year. Deltas are situated in strongly subsiding areas.

As the delta builds out, the pro-delta muds are progressively covered by the coarser sediments of the delta front and then of the alluvial plain. Thus the sedimentary succession shows a general **vertical increase in grain size**.

b) The Organisms

In the subaerial part of the delta, freshwater and terrestrial organisms predominate. The many stands of **freshwater** encourage an abundant vegetation, whose roots may be fossilised in situ, and which can give rise to coal and lignite.

A mixture of fresh and salt water occurs in the submerged part of the delta. The fauna is characteristic of **brackish environments**. When the sediment is fine-grained, it is heavily bioturbated by burrowing organisms. The high rate of sedimentation on the delta front does not allow the establishment of lasting communities.

Seawards, the pro-delta muds carry a rich benthic marine (or lacustrine) fauna dominated by burrowers; they also contain pelagic remains.

2. **Estuaries**

Being under the influence of the tides, estuaries have many features in common with the intertidal zone. The alternation of ebb and flood tides produces opposing palaeocurrent directions in the deposits.

The sediments are generally fine-grained; sands are localised in channels while mud accumulates on the banks where it is easily fixed by plants. This mud is frequently ripped up as soft **mud pebbles**, giving rise to true intraformational conglomerates.

The fauna consists of **brackish water** species. Upstream they are replaced by freshwater forms and downstream by a marine fauna.

IV. Lagoons

Lagoons are bodies of water in temporary or permanent communication with the sea, from which they are separated by a natural barrier. This is a low energy environment which, since the water is shallow, is subject to a great variation in physico-chemical parameters: temperature, salinity, oxygenation, etc.

1. The Sediments

Lagoonal deposits are fine-grained, including sands and muds which are well bedded but often bioturbated. They are often rich in organic matter, especially **faecal pellets**. There may be signs of rhythmic deposition caused by climatic variations. Ripple marks are common.

When terrigenous input is low, carbonate precipitation (micrite) becomes dominant. In an arid climate, evaporites may be deposited.

2. The Organisms

The biological content, low compared to that of the neighbouring sea, faithfully reflects the physico-chemical conditions of the lagoonal environment. Sometimes the fauna is marine or, more often, typical of **brackish water** (molluscs, crustacea, insects). Burrowers are abundant. Occasionally the aquatic fauna suffers **mass mortality** because of more rigorous climatic conditions (warming of the water, isolation, drying out, etc.). When the salinity becomes too high, all life disappears and the sediments are barren.

The gradual filling-up of the lagoon assists colonisation by plants, such as mangroves.

V. Turbidity Currents

Turbidity currents are density currents, laden with material in suspension, which flow under gravity and drop their load in the deep parts of basins.

They are very swift phenomena which occur close to an underwater slope (1° is sufficient) following a breakdown of equilibrium which sets off mass movement of sediment. As the turbidity current flows, it is transformed into a mass of mud denser than the surrounding water and is able to erode the underlying sediments. As the current slows, its transporting ability decreases and particles are dropped in decreasing order of grain-size: thus a horizontal and vertical grading is produced. The resulting deposit is called a **turbidite**.

The sedimentary process which produces turbidities can be used to explain the emplacement of flysch in geosynclines. Actually, it is known that turbidity currents can develop in very varied environments and deposit sediment even in shallow water.

1. The Sediments

Each turbidite represents a **sedimentary sequence** (Fig. 85) whose base is an **erosion surface** covered with sedimentary structures (current struc-tures – **flute casts**). The fining-upwards grading is very marked. At all levels in the turbidite, the coarse grains are held in a fine-grained matrix. Above this, planar parallel laminae appear. In the fine-grained parts of the unit, these laminae are deformed by slumping or compaction. The top passes gradually to the normal deep-sea sediments; this is Lombard's "background noise".

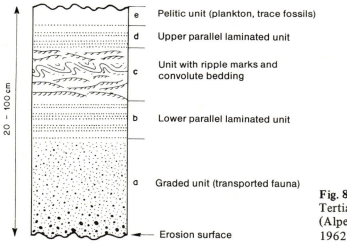

e — Pelitic unit (plankton, trace fossils)

d — Upper parallel laminated unit

c — Unit with ripple marks and convolute bedding

b — Lower parallel laminated unit

a — Graded unit (transported fauna)

— Erosion surface

20 – 100 cm

Fig. 85. Ideal sequence of the Tertiary flysch at Peira-Cava (Alpes-maritimes). (Bouma 1962)

These sequences often cover a large horizontal area. Close to their source, they can occur in channels (submarine canyons). Moreover, as one passes from proximal to distal deposits, one can see a change from thick, irregularly-bedded, coarse sediments with signs of erosion, to fine-grained, thinly laminated sediments with few sedimentary structures. Palaeocurrent measurements can give the direction of the flow and, in certain cases, the palaeotopography of the basin.

2. The Organisms

The lower part of the sequence contains organisms carried with the sediment from shallow water. They are often benthic forms and are graded just like the sediment.

The pelitic part of the sequence contains the tests of plankton (foraminifera, radiolaria, diatoms, etc.) and the remains of pelagic animals (fish teeth). In flysch, these levels often contain a varied and well-preserved **ichnofauna**: grazing traces, feeding burrows, etc.

VI. Environments of Deposition of Evaporites

Saline deposits (rock salt, gypsum, sylvite, etc.) can form in marine environments or on a continent.

At the present time, it seems that evaporites in continental basins have been largely derived from older salt deposits. Most of the large occurrences of salt were originally deposited in a marine environment.

Two conditions are required for their formation:
— a **dry climate** leading to active evaporation;
— a **great diminution of detrital input.**

The depth of water is not a determining factor in their formation.

There are several processes by which the precipitation and accumulation of evaporites may be explained.

1. Lagoons

When a lagoon is partially separated from the main body of the sea by a **bar** (Fig. 86), intense evaporation will concentrate the water. Its density increases and it can accumulate in the deep parts of the lagoon where salts are deposited. When the surface water is replenished by the sea, the process can continue for a long time.

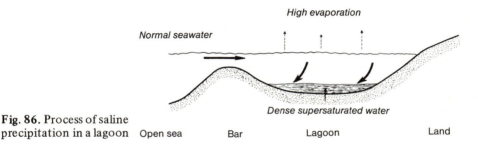

Fig. 86. Process of saline precipitation in a lagoon

2. Major Marine Platforms

In this case, the process of precipitation of the salts is similar to that in lagoons; however, the return current of salt-laden water is not restrained by a physical object (the bar) but by a moving obstacle. Effectively, the differences in density of the waters and the friction at the interface of the two fluids, etc., impede the mixing of the water and the return of the the highly salt brine. On a marine platform, there is thus a salinity gradient between the areas closest to the open sea where the least soluble salts are deposited (gypsum) and the areas closer to the landmass where the most soluble salts are precipitated (rock salt and sometimes potassium salts).

Busson has described the formation of the Triassic and Jurassic of the Sahara platform in this way.

Salts deposited by these two processes are thick, well-bedded and of great lateral extent. They are often associated with dolomites and are usually lacking in fossils.

3. Precipitation Within the Sediment

The salts originally contained in the depositional water of marine muds can crystallise after burial. These evaporites form irregular pockets or isolated crystals. According to Bourcart and Ricour, some salts within the French Trias formed in this way.

In the **supralittoral zone**, dolomite, anhydrite, gypsum or rock salt can form in the sediment by diagenetic replacement. The supply of brine occurs by capillary action and by temporary flooding. Such evaporites are of little importance in nature at the present time, but have been identified in several fossil formations.

Chapter 6 **Fossiliferous Horizons**

Fossiliferous horizons are sedimentary units which are particularly rich in palaeontological evidence, both because of their diversity of species and the exceptional state of preservation of the fossils. Fossil hunters find them very attractive, and for the geologist they are an unequalled source of information on ancient environments and can give great assistance in interpreting deposits of the same age which are less rich in fossils.

These fossiliferous horizons are the result of a long series of biological and physico-chemical changes which have considerably altered the appearance of the original biotopes.

The task of the palaeoecologist is to reconstruct these fossil environments and their populations despite these changes.

This process has several stages:

— description of the methods of fossilisation;
— determination of the cause of death of the organisms and of their accumulation;
— reconstruction of the original biological associations and their life habits and habitats.

I. Origin of Fossiliferous Horizons — Taphonomy

Taphonomy is the post mortem history of organisms, in other words, the sequence of events that has taken place between the death of the organism and its present day state of fossilisation.

1. Accumulation of Organisms

Many fossiliferous horizons are characterised by a high concentration of fossils. This can have different origins.

a) Dense Populations

Very dense populations of organisms, where individuals grow on one another, can give rise to organic build-ups whose size depends on the length of time they are active; reefs and peat bogs are examples.

b) Mass Mortality

Mass mortality of populations can occur when there is a sudden and violent change in the physico-chemical character of the environment. Hurricanes, floods or volcanic eruptions are important causes of death in land organisms. In water, the arrival of turbid waters can cause the asphyxiation and burial of many benthic forms. The release of hydrogen sulphide or the seasonal proliferation of toxin-producing plankton (**blooms** of blue-green algae, dinoflagellates, etc.) can also exterminate whole faunas. When bodies of water dry out, aquatic or amphibious animals are concentrated in the residual pools as they die.

c) Transport

Organisms, often coming from separate biotopes, can be concentrated and trapped by transport, whether short or long. Shell beds and most coal horizons are formed in this way.

2. Burial of Organisms

After death, all organisms are affected by physico-chemical and biological processes which begin their destruction.

— **In an oxygenated environment**, the soft parts decay. Saprophytic and necrophagous organisms, especially bacteria which cause decomposition, are particularly important in this. The shells of brachiopods and bivalves, the tests of echinoderms, the skeletons of vertebrates, etc., are disarticulated mechanically. The bodies of vertebrates will float on the surface of the water for a certain amount of time, buoyed up by gases from putrefaction which accumulate in their body cavities. As they drift, successive parts of the skeleton become detached and drop to the bottom, resulting in incomplete skeletons and separated bones. Finally, currents separate the hard parts and disperse their products. Over a period of time, these different processes can lead to the complete disappearance of the bodies.

In the absence of water, bodies are **mummified**.

— **In the absence of oxygen,** the action of these different processes is considerably slowed down. Plant remains are carbonised; the decomposition of animals produces **sapropels**. These conditions occur when the bodies are rapidly buried in fine-grained sediment (clay, calcareous mud, etc.) where oxygen/water exchange is low. This explains the bacteria preserved in Palaeozoic crystals of rock salt and the arthopods in the Baltic Oligocene amber as well as the mammoths frozen in the Quaternary ice in Siberia. The fine-grained matrix also faithfully reproduces the morphology of the fossils which are preserved as **moulds** when the original material has entirely gone.

To summarise, the ideal conditions for fossilisation are burial in fine-grained sediment in the absence of oxygen.

3. Diagenesis

After burial, organic remains undergo a slow and gentle change by the exchange of materials. Exactly what these changes are depends on the nature and porosity of the surrounding sediment, as well as on the chemical composition of the fossils themselves and their decomposition products. All these reactions are part of diagenesis, which results in a replacement or induration of the hard parts by recrystallisation or mineralisation.

In general, limestones favour the preservation of carbonate shells, carapaces and skeletons, which tend to be dissolved in sandstones. There, however, fossils can be **silicified** which preserves minute anatomic detail (for example: silicified wood). **Dolomitisation** of carbonate remains generally alters their original structure. In clays and marls, **concretions** of various minerals (limestone, siderite, etc.) frequently form around fossils. When organic matter is abundant, free sulphur induces **pyritisation** of fossils.

Fossiliferous horizons are the result of a great number of processes which take place concurrently or successively in the course of their formation. Every situation requires its own approach; several examples of this will be described in Part Two of this book.

4. Classification of Fossiliferous Horizons

a) Horizons Formed by Concentration

Some fossiliferous horizons are the result of the concentration of the hard parts of various organisms. The skeletal elements are disarticulated (Fig. 87). These horizons form where there is a low rate of sedimentation (for example: silts in Quaternary caves). They can also accumulate by transport in a current which has sorted biogenic fragments (for example: shell beds, crinoidal limestones (Pl. I, Fig. 2), **bonebeds** made of vertebrate

Fig. 87. Fossiliferous horizon formed by the concentration of organisms: the fossils (valves, carapaces, bones, etc.) are disarticulated and broken. It is cross-bedded. (Schäfer 1962)

Fig. 88. Fossiliferous horizon formed by the preservation of the organisms: the fossils (shells, skeletons, etc.) are autochthonous and often whole. Animal trails and burrows are abundant. Bedding is horizontal. (Schäfer 1962)

bones and teeth, etc.) or which has collected organic remains in a sedimentary trap (for example: fissure infillings).

b) Horizons Formed by Preservation

This second category of fossiliferous horizon is characterised by the partial preservation of organic matter (especially chitin) and by the presence of tests or skeletons which are still articulated (Fig. 88). Stagnant bodies of water where conditions are anaerobic favour this type of fossilisation (for example: bituminous shales, lithographic limestones, etc.). The same result is obtained when the bodies are rapidly buried by a reducing sediment or in a preserving medium (for example: amber) or preserved in some early diagenetic concretions (for example: chalk flints).

II.   Associations of Organisms – Palaeosynecology

Even in optimal conditions, only a small proportion of the original population is fossilised. Soft-bodied animals and most micro-organisms are not usually found by palaeontologists. As an estimate, in a present-day biotope **less than** 1% of the species is likely to be fossilised. Moreover, the specimens collected come primarily from the benthic population. The sheer size of this **loss of information** means that the concepts and methods of ecology cannot be directly applied to the reconstruction of ancient environments. Specifically, since the concept of the **ecosystem** is analogous to that of **facies**, a floral and faunal list from the fossiliferous horizon can never be the equivalent of a community of living organisms. Many groups have not been fossilised and the forms listed might have come from successive populations or from completely different biotopes.

These difficulties are partially eased by detailed study of the sediments and organisms in the fossiliferous horizons. When presented with associations of fossils, the palaeoecologist must answer this question: did the species which have been collected live where they were buried or are they an accidental collection of forms from different biotopes and different ages?

1. Palaeobiocoenoses

With the above reservations, associations of fossils which lived together at any one moment in the place where they were buried can be considered to be biocoenoses.

Such conclusions can be drawn when the organisms are in their **position of life**. Examples are reefs built up by the carbonate skeletons of generations of encrusting organisms, and some fossil swamps where the root systems are preserved in situ in a palaeosol (Fig. 124). Fixed or burrowing benthic species (brachiopods, bivalves, etc.) can also be found in the relationship to the substrate which they had when alive (hardground populations, for instance). Isolated examples of fossil relationships, commensualism or parasitism, have been repeatedly described, for example, the classic association of the Devonian tabulate, *Pleurodictyum,* and a worm, *Hicetes* (sipunculid?), which is interpreted as **symbiosis** benefitting the latter (Fig. 89). However, the best evidence of autochthonous life is provided by traces of animal activity (trails, burrows, etc.).

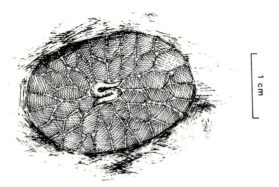

1 cm

Fig. 89. Symbiosis between *Pleuro-dictyum,* Devonian tabulate coral, and a worm *(Hicetes)*

Many fossil horizons formed by conservation yield palaeobiocoenoses. Various observations can be used to deduce that the organisms are in place:
— absence of signs of transport and mechanical sorting;
— completeness of skeletons (for example: the two valves of bivalves still articulated);
— the co-existence of several stages of development in a monospecific population;
— an abundance of juveniles resulting from a high mortality shortly after hatching;
— the presence of signs of biological activity: eggs, coprolites, trails, burrows, etc.

Organisms in their life position and the nature of the sediment (fineness of sedimentary grains, absence of current marks and structures, etc.) will confirm that the horizon being studied represents an original habitat whose populations have been little reworked.

2. Thanatocoenoses

Thanatocoenoses, which have also been called **taphocoenoses** (Quenstedt) or **symmigies** (Babin), are the result of the mechanical accumulation of organisms and are the cause of fossiliferous horizons formed by concentration. They do not represent a living community but the result of a long post mortem series of events for the organisms.

The fossils in a thanatocoenosis are allochthonous. They can come from varied environments and geological times. It is the job of the palaeoecologist to distinguish those parts of the association which come from an original biocoenosis from those which are the results of taphonomy. Both the characteristics of the sediment (indications of a high energy environment, sedimentary breaks, etc.) and the appearance and position of the fossils within a horizon can help in deciding the degree of derivation of an association.

a) Orientation of Fossils

During their transport by water or by wind, dead organisms behave like sedimentary particles (Fig. 119). Under the influence of a unidirectional current, elongated shells (gastropods, orthocones, belemnites, etc.) lie with their long axis parallel to the direction of flow (Fig. 90). In the majority of cases, the sharp end will point upstream. On the other hand, if the heavier end is anchored in the sediment, the sharp end of the shell will point downstream. In the intertidal environment, the ebb and flow of the waves will arrange the long axes at right angles to their direction

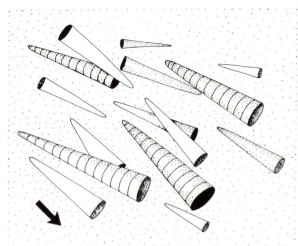

Fig. 90. Current orientation of Silurian orthocone shells

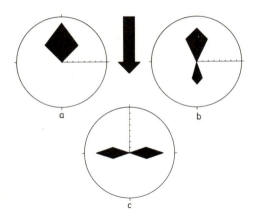

Fig. 91. Orientation of the anterior-
posterior direction of animals coming
from the same horizon. a in life position (suspension feeders); b affected
by a current; c affected by waves.
Arrow indicates the direction of the
current or the movement of the waves.
(Modified from Seilacher 1960)

◀ Fig. 92A, B. Valves of a bivalve: A in quiet water (unstable equilibrium); B in active water

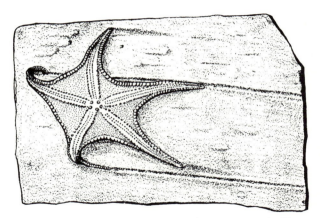

Fig. 93. Orientation of a starfish *(Euzonosoma)* affected by a current in the Devonian Hunsrück Slates.
Direction of the current *from right to left*. (Seilacher
1960)

of movement and hence the shells will accumulate in the troughs of
ripples. In both cases, the orientation of the fossils differs from that of
suspension feeders whose openings always face the current when they are
in their life position (Fig. 91).

When they lie concave-up, the arched valves of brachiopods and bivalves
are in unstable equilibrium on the substrate and are easily overturned
even by a weak current. Valves which have been transported are buried
convex-up in the sediment (Figs. 92, 93).

b) Sorting by Size and Weight

During transport, skeletal parts are sorted by size and weight. Many accumulations of fossils are made up of organisms or fragments of hard parts which are noticeably similar in size (bone beds, crinoidal limestones, some shell beds, etc.) (Pl. I, Fig. 2).

In associations of bivalves, sorting can appear as a higher proportion of right or left valves. The two valves often differ in their shape and in the distribution of teeth on the hinge and this affects their hydrodynamic behaviour. In the Muschelkalk, for example, there are often three times as many left valves as right valves. Sorting of fossils can be easily seen in size-frequency curves (Fig. 94).

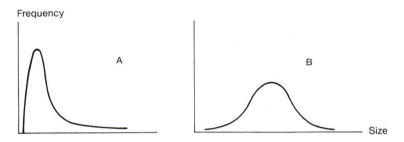

Fig. 94A, B. Autochthonous (biocoenosis) or allochthonous (thanatocoenosis) origin of an association of animals deduced from frequency curves. **A** Palaeobiocoenosis; **B** Thanatocoenosis. (Boucot 1953)

c) State of Preservation

Transport and successive reworking of fossils can cause a marked deterioration of the hard parts. Fragile, branching shapes and juveniles disappear first. Shells, carapaces and skeletons are disarticulated, broken and worn (Fig. 95). The angles of skeletal fragments become rounded. This mechanical action is usually helped by chemical (partial solution of shells in an acid environment) and biological (boring organisms: algae, fungae, sponges etc.) factors.

Fig. 95. *Patella* shell worn by prolonged transport

d) Indicators of Recycling

A thanatocoenosis is rarely the result of a single cycle of transport.
Before they are finally buried, fossils are often disinterred by erosion and
redeposited elsewhere. This is especially clearly seen when the infilling of
shells differs from the matrix.

Similarly, the sparite cement which forms in the half-empty shell
above a partial fill of sediment (geopetal infilling) is also an indicator.
The boundary between the two infills represents the horizontal when the
shell was first fossilised (Fig. 96; Pl. I, Fig. 1). After reworking, its orientation with respect to bedding betrays the fact that the specimen has
been recycled.

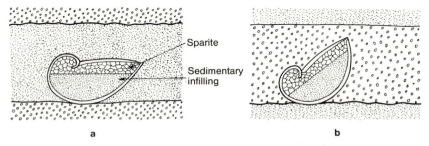

Fig. 96a, b. Geopetal infilling in fossil shells. a shell in original position; b reworked shell

e) The Mixture of Floras and Faunas

When thanatocoenoses consist of organisms which quite clearly belong
to different biotopes (for example: mixing of hard-bottom benthos with
burrowing species), it is easy to tell that they are allochthonous.

III. The Study of Fossiliferous Horizons

1. Uniqueness of Palaeoecological Methods

The methods of analysis used by ecologists cannot be directly applied to
the study of fossiliferous horizons, in which one is conscious of the successive populations which colonised the environment or which immi-
grated at different times. The existence of this time dimension means
that the palaeoecologist must dissect a horizon level by level. On the geological time scale, these levels represent instants and may sometimes consist of laminae a few fractions of a millimetre thick (Fig. 97).

Like the archaeologist, the geologist faces a fossilised world. The characteristics of the original environment are accessible to him only through the organisms and the sediments, and he is forced to unravel it **inductively**. This is another reason why palaeoecological studies are different.

2. Collection of Information

Fossils can be **sampled qualitatively and quantitatively**. When studying fossiliferous horizons, one must consider all the organisms, whole and fragmentary, as well as the signs of their activity. Only a complete census of every organism will produce a satisfactory picture of the original population. The relative abundance of different species should appear on faunal and floral lists so that the dominant species, the auxiliary species and ones which are merely there by accident can be identified (Fig. 24).

Special attention should be paid to the **state of preservation** of the fossils and their spatial **position**. These observations often allow one to distinguish between biocoenoses and thanatocoenoses.

Study of the **enclosing sediment** will provide a great deal of evidence about the physico-chemical characteristics of the environment and its place in the regional palaeogeographic picture.

3. Presentation of Results: Palaeoecological Profiles

All palaeobiological and sedimentological information collected from a fossiliferous horizon is eventually synthesised, level by level, on palaeo-ecological diagrams or profiles (Fig. 97). These allow one to grasp the sedimentary history and the succession of floras and faunas through time at a single glance. From the juxtaposition of different pieces of information, it is possible to work out the essential features of the environment (Fig. 98, Pl. II, Fig. 1).

Fossiliferous horizons are thus windows which throw a clear but pinpoint light on the floras and faunas which have followed each other through the night of time. Their study calls on various disciplines: palaeontology, sedimentology, geochemistry, etc., and this is ideal for a team. The specialist workers on prehistoric sites are particularly distinguished in this field.

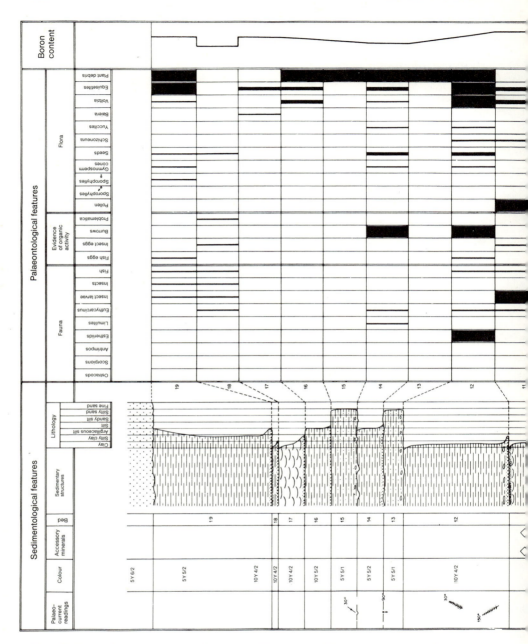

Fig. 97. Example of a palaeogeological profile of a fossiliferous horizon in the Bunter Sandstone (Grès à Voltzia) of the Vosges. (Gall 1971)

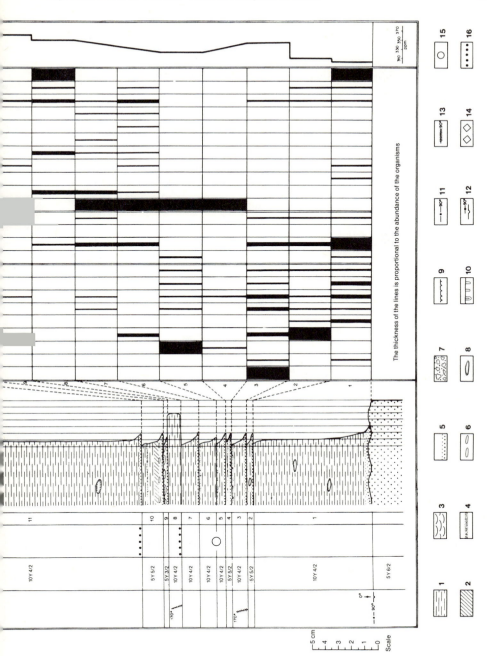

Fig. 98. Reconstruction of a Carboniferous landscape in the north of France. *1 Lepidodendron; 2 Lepidodendron or young Ulodendron; 3 Ulodendron; 4 Lepidophlois; 5 Sigillaria; 6 young Sigillaria; 7 Sphenophyllum; 8 young Asterocalamites; 9 young Calamites; 10 Cordaites.* (P. and P. Corsin 1970)

Part Two
Reconstruction of Some Ancient Environments

Chapter 7   The Ediacara Fauna

Although it occurs outside Europe, the Ediacara fauna will be described here because it represents a fundamental stage in the history of life. As far as we know at present, it is the oldest example of a biotope.

The Ediacara hills are in southern Australia, between Lake Torrens and the Flinders Ranges, 600 km north of Adelaide. The fossil horizons can be followed laterally for 140 km. Some parts of the fauna have been recorded from other horizons elsewhere in the world and dated as being between 600 and 700 million years old, close to the end of the Precambrian.

I. The Sediment

The Ediacara fossiliferous horizons are divided into two main units, each some tens of metres thick. They consist of fine sandstones, siltstones and shales which have a reddened surface. They are interbedded with unfossiliferous horizons of flat bedded or cross bedded feldspathic sandstones. Their base and top are gradational.

The fossiliferous horizons are subhorizontally planar parallel bedded. Individual laminae are often irregular and are grouped into thicker units separated by shaly or silty horizons, or small lenses of sandstone. The tops of the units are often **ripple-marked** and sometimes truncated by a penecontemporaneous erosion surface. The ripple-marks and occasional scour structures are the only evidence for local currents. Large scale channels appear only at the base and top of the fossiliferous horizons.

There is a gradual change in sedimentation through the sequence: from bottom to top, sorting improves, the sandstone horizons become thicker, ripple marks become more frequent and silts and shales become less important. This indicates a gradual increase in hydrodynamic energy.

II. The Fossils

The interpretation of Precambrian fossils is fraught with difficulty. Firstly, pre-Palaeozoic organisms are very different from present-day forms or are still at a very early stage of evolution. Their simple morphology can easily be confused with inorganic structures. Secondly, Precambrian sediments have had a long and complex geological history which has frequently altered any traces of flora and fauna. The demonstration of different ontogentic stages and the occurrence of trails and burrows are thus the only criteria for attributing an organic origin to these remains.

As far as we know at present, the first signs of life on earth date from more than 3000 million years ago (early Precambrian) in the form of bacteria and unicelluar algae. The first metazoans did not appear until late Precambrian times.

In 1971, 1600 specimens were collected from the Ediacara horizons. They included coelenterates (67% of the total number of specimens), annelids (25%), arthropods (5%) and some forms whose affinities are still doubtful (Fig. 99).

Fig. 99. Reconstruction of the Precambrian Ediacara environment. Coelenterates: *1 Ediacara; 2 Mawsonites; 3 Rangea; 4 Pteridinium.* Annelids: *5 Spriggina; 6 Dickinsonia.* Arthropods: *7 Parvancorina.* ?: *8 Tribrachidium; 9* Algae? (Modified from Glaessner 1971)

1. The Fauna

a) Benthic Organisms

▫ **Coelenterates**
Rangea and *Pteridinium* are related to the pennatulid group of alcyo-narians. The polyp colonies are held on leaf-like sheets supported by a fixed stem. They are locally rather abundant in these horizons, which suggests that they lived close to the place where they were buried.

▫ **Annelids**
Segmented forms have been attributed to the annelids. *Dickinsonia* had a flat, oval body in which the trace of the digestive tract can be seen. It was probably microphagous and must have lived in a neigh-bouring, muddier biotope. *Spriggina,* which was more slender, shows a greater morphological differentiation, closer to that of the free-living polychaetes.

▫ **Arthropods**
Parvancorina resembles the larva of a very large trilobite. The Ediacara fauna contains examples at different stages of their ontogeny, which proves that they lived in the locality.

▫ **Forms of Uncertain Affinities**
Tribrachidium is a disc-shaped organism of which the face which is presumed to be oral carries tri-radiate arms. This is reminiscent of some echinoderms, but the body was not strengthened by calcareous plates.

b) Nektonic Organisms

About ten species of **medusid** have been found at Ediacara. Their bells, which are several centimetres in diameter, are decorated with furrows or lobes and sometimes show marginal tentacles. They are related to the scyphozoans and the hydrozoans. Some particularly abundant species *(Ediacara)* seem to have led a gregarious life.

   None of the organisms of the Ediacara fauna had a mineralised exo-skeleton.

2. The Flora

Some problematical structures have been attributed to the **algae**.

3. Evidence of Biological Activity

Some traces of animal activity have been described from the Ediacara fossiliferous horizons, chiefly gallery systems excavated by sediment feeders, probably worms, and grazing meanders left on the surface of the sediment by detritus feeders. These traces indicate that the original sediment must have been fairly rich in organic matter.

III. Preservation

The Ediacara fauna is preserved as moulds or imprints. Nothing remains of the original organic material. The state of preservation is at its best when the fossils lie at the junction of two beds of differing grain size and when the overlying bed is less plastic than the underlying bed.

Much can be learnt about the speed of burial of the medusids from their preservation. In normally-oxygenated water, medusids are reduced to a thin gelatinous disc in about ten days. However, some Ediacara medusids show that several laminae were deposited before the decomposition of the

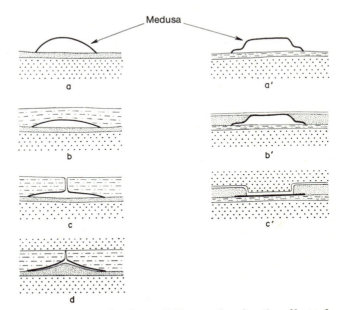

Fig. 100a–d. Two aspects of medusid fossilisation at Ediacara, showing the effect of a high rate of sedimentation. **a–d** decomposition of the animal occurred after burial and caused the formation of an escape channel for the gas produced; **a'–c'** decomposition of the animal occurred after burial, causing the overlying laminae to subside. (Modified from Wade 1968)

animals allowed the sediment to subside. Moreover, some examples are related to vertical channels which correspond to the line taken through the still-unconsolidated sediment by decomposition gases from the medusids (Fig. 100). These observations indicate **a high rate of sedimentation**.

Most of the medusids, however, have been fossilised with the convex part of the bell downwards. This orientation is the opposite to that normally observed today in medusids stranded on beaches. It could only occur if the depth of water was greater than a limiting value equivalent to about two-thirds of the diameter of the animals. This suggests that, at Ediacara, fossilisation took place in a permanent sheet of water. This conclusion is supported by the lack of traces of emergence.

To summarise, in the Ediacara horizons, the preservation of soft-bodied animals may be explained by rapid burial in a fine-grained sediment.

IV. The Environment

The gradual passage which can be observed between the fossiliferous horizons and the planar or lenticular bedded surrounding sandstones at Ediacara indicates that the two facies were deposited within the same palaeogeographic setting. The sandstones represent an environment of marine sedimentation with strong currents. In the fossiliferous horizons, the gradual increase in frequency of ripple marks and the progressive decrease in the fine fraction of the sediment with time are related to a rise in energy level corresponding to a reduction in water depth. Some authors have estimated this to be 25 m. These depths are within the **neritic zone,** but it is not possible to state the distance from shore.

However, the oscillation ripples, the fineness of the sediment and the low frequency of erosion surfaces within the laminated horizons suggest only a moderate movement of the water. The oxygenation of the sediment was sufficient to allow benthic and burrowing forms to live there, but the fauna is too unlike anything else to make further deductions about the conditions of life.

The environmental picture that appears is one of a bay temporarily sheltered from the sea by a barrier which formed an obstacle to currents. The medusids which had been washed in from the sea found optimum conditions for fossilisation because of the high rate of sedimentation.

The Ediacara fossil horizons have thrown light on one of the oldest environments in the history of the earth, that of a shoreline region populated by a varied community of simple, soft-bodied animals.

Chapter 8   The Old Red Sandstone Continent

In **Devonian** times, the erosion of the Caledonian chain gave rise to several thousands of metres of brightly coloured conglomerates, sandstones and shales. These detrital sediments are called the Old Red Sandstone. They accumulated on a vast North Atlantic continent stretching from the Rocky Mountains to the Urals, with especially fine outcrops in Scotland and Wales.

There are many environments within the Old Red Sandstone. Three of them are described here: the fluviatile complex, the lacustrine deposits and the peat bogs.

I. The Fluviatile Complex

Most of the detrital material of the Old Red Sandstone has been deposited by a fluviatile complex. This sedimentary process can be studied in greatest detail in Wales and the Welsh Borders (Shropshire, Herefordshire and Worcestershire) of England. The **Clee Hills**, north of Ludlow (Shropshire), are considered to be a classic area; there the Lower Devonian rocks belong to the **Dittonian** stage.

1. The Sediment

a) Petrology

The Dittonian is about 400 m thick in the region of the Clee Hills; it consists of a complete range of detrital sediments from conglomerates and sandstones to siltstones and has calcareous horizons. The grains are angular and moderately well rounded. Accompanying quartz are fragments of volcanics and older sediments which indicate possible source areas. Because of the presence of iron oxides, the dominant colour of the rocks is red.

b) The Cyclothems

Bedding is typically lenticular and the coarse parts of the rock are generally cross-bedded.

The apparent heterogeneity of the rocks can be rationalised into a succession of lithological sequences or **cyclothems**, repeated many times. Each of the cyclothems, whose thicknesses vary between 3 and 18 m, is made up of the following units, from bottom to top (Fig. 101):

– a **conglomerate**, mainly derived from the underlying rocks and separated from them by a strong erision surface with scour marks; it is cross-bedded;

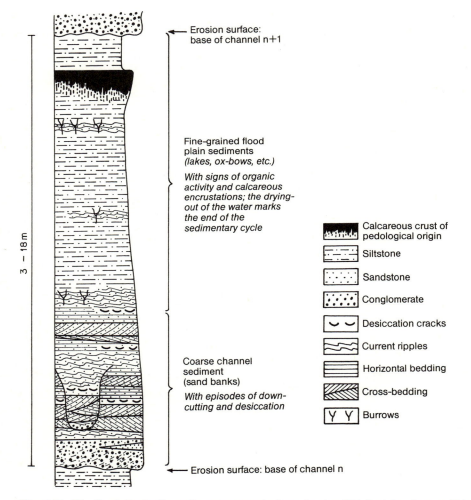

Fig. 101. Idealised fluviatile sedimentary cyclothem in the Old Red Sandstone (Dittonian) of Wales. Thickness is exaggerated. (Allen and Tarlo 1963)

— cross-bedded **sandstones** with current ripples and frequent channels, whose fill may be up to ten metres thick. They are interbedded with fine-grained, well-sorted, horizontally bedded sandstones which show a parting lineation. Locally there are desiccation cracks. The sediment fines upwards and current ripples become more frequent;
— **siltstones** without apparent bedding but some levels of current ripple marks; carbonate nodules are abundant towards the top;
— the cyclothem is topped by a **carbonate crust**.

The grain size diminishes upwards in the sequence; the sedimentary structures indicate a current regime which is becoming weaker and less competent with time. These changes, together with the consistency of palaeocurrent directions, are typical of a fluviatile regime.

Fossiliferous lenses of green fine-grained sandstone sometimes occur between the conglomerate and the sandstones.

2. The Organisms

The fossils collected in the Dittonian of the Clee Hills are chiefly of aquatic vertebrates: agnathans and fish. Some levels also yield molluscs, arthropods and plant remains. These fossils are usually found at very well defined levels within the cyclothems.

a) Vertebrates

Ostracoderms are agnathans whose flattened body is protected by a heavy, bony armour. Fragments belonging to different genera are found in the conglomerates at the base of the cyclothems *(Cephalaspis, Psammosteus)*. The lenses of finer sediment interbedded between the conglomerates and the sandstones have yielded whole specimens, associated with bony fish (acanthodians).

b) The Invertebrates

The remains of **eurypterids**, which were large carnivorous merostomes, have been found with the vertebrates. Burrows occur in the siltstones.

In the Lower Dittonian, where there is still some marine influence, the finer units of the cyclothems have yielded bivalves *(Modiolopsis)*, gastropods and ostracodes *(Leperditia)*.

c) The Flora

Plant remains, which are difficult to identify, are mainly concentrated in the finer deposits. Moreover, the calcareous crusts at the top of the cyclothems have been thought to have a **pedological origin**, as calcareous concretions formed around root systems in a hot, dry climate.

3. Fossilisation

Most of the vertebrate remains consist of fragments which have suffered long transport before burial. They come from the regions upstream of the deposit. On the other hand, the more or less complete skeletons found in the lenses of fine sediment have hardly been moved any distance after death. Different growth stages of an agnathan can be found in the same horizon. Elsewhere, an acanthodian still has the carapace of an ostracoderm, which it had swallowed, in its stomach. These animals were fossilised close to where they lived. The fine grain of the sediment suggests relatively quiet water. Its green colour is probably due to the reduction of ferric oxide, which gives the normal red colouration, as a result of the decomposition of plant remains.

Protected from active channels, in poorly oxygenated basins with a high rate of sedimentation, these animals found suitable conditions for fossilisation.

4. The Environment

The lenticular bedding, the vertical changes in the cyclothems and the constancy of palaeocurrent directions are characteristic of fluviatile sedimentation. Each sequence represents the brief history of a watercourse. During a flood or a break in the natural embankments, a mass of water laden with detritus suddenly spread over the lower parts of the alluvial plain. It eroded the substrate and cut a channel in it. In the course of time, the carrying power of the water grew less and it gradually dropped its load. First it dumped the coarse material, rich in pebbles which it had picked up as it passed, mixed with plant remains and disarticulated fragments of the vertebrates which had lived upstream. When the environment became less turbulent, sands were deposited and lower energy structures such as current ripples became more common. Locally there were lakes with quieter water where agnathans, fish and eurypterids lived for some time before dying because of the cutting-off and drying-out of the water. A fall in water level or lateral migration of the channel occurred and a lower energy regime allowed deposition of finer sediments, the siltstones. The mud was then colonised by burrowing organisms. In the seaward part of the flood plain, occasional marine incursions took place.

As it evaporated, the body of water became more concentrated and carbonates were precipitated. The invasion of swamp vegetation assisted the fixing of limestone around its roots. This marked the end of the fluviatile sequence.

With the breaking of an embankment or another flood, the channel took a new course and a new cycle began.

In early Devonian times, the south of England was thus a huge **alluvial plain** close to the sea. Regional subsidence and the lateral migration of watercourses over millions of years led to the deposition of detrital sediments washed down from the neighbouring mountains. Bodies of freshwater were populated by agnathans and fish who might have first evolved in the sea since they are so widely distributed geographically.

II. The Orcadian Lake

Achanarras quarry, in Caithness in the extreme northeast corner of Scotland, has been known for more than a century for its rich horizons of Middle Devonian Old Red Sandstone fish.

1. The Sediment

a) Petrology

About 10 metres of hard, grey, fissile flagstones, excellent for paving and roofing, are worked at Achanarras quarry. Petrologically, the rock is a siltstone with a high carbonate content (calcite and dolomite). It contains organic matter and small grains of pyrite. The rusty colour which sometimes develops at outcrop is due to the presence of iron carbonate.

b) Stratinomy

i) Bedding

The Achanarras flags outcrop in regular units a few decimetres thick, with a well-marked finer-scale bedding. The laminae, some tens of millimetres thick, are separated by horizons rich in organic matter. This distribution makes the rock very easy to split.

Sometimes the laminae adopt a rhythmic pattern reminiscent of varves: their base is rich in carbonates while their top mainly consists of quartz grains and most of the organic matter. The detrital fraction is graded.

ii) Sedimentary Structures

The only sedimentary structures to occur within the flags are **desiccation cracks**, *"anti-ripplets"* produced by dry grains sticking to a wet sandy surface as they are blown over it by the wind, and *"rill marks"* formed by trickling water. These all indicate emergence.

2. The Fossils

a) The Fauna

The Achanarras horizon chiefly contains fish. About ten genera belonging to the placoderms, the acanthodians, the dipneusts, the crossopterygians and the actinopterygians have been listed (Fig. 102). The spatial density of the fossils is important: one can collect five or six fish on a 1 m² surface. The **dipneusts** *(Dipterus)* and the **placoderms** *(Coccosteus)* are especially abundant.

The former are characterised by a lung which enables them to survive in stagnant water by taking oxygen directly from the air. The latter are the "armoured fish", related to the present day holocephalians (chimerids). The anterior part of their body is protected by a strong bony armour articulated at head level. This armour limited their movements and they probably moved about in contact with the bottom. *Coccosteus,* which had cutting dental plates, and the actinopterygians were carnivores. Doubtless they ate young fish or invertebrates whose remains have not been preserved. Aquatic plants must have formed part of the diet of the dipneusts.

b) The Flora

Apart from a few spores, the plant remains of Achanarras are unidentifiable. Phytoplankton are probably responsible for the high organic content of certain horizons.

c) Evidence of Biological Activity

Rare disturbances of the surface of the flags have been interpreted as traces left by burrowing organisms.

Fig. 102. Agnathans *(1, 2, 3)* and armoured fish *(4)* from the Old Red Sandstone of England. *1 Hemicyclaspis; 2 Poraspis; 3 Pteraspis; 4 Coccosteus.* (Piveteau 1951)

3. Preservation

The fineness of the sediment and the regularity of the bedding suggest that deposition took place beneath a body of quiet water. The absence of currents can also be deduced from the excellent preservation of the fish whose scales and bones are thinly scattered on the surface of the flags.

The rarity of burrowers, together with the richness in organic matter and pyrite, indicate that **reducing conditions** developed in the bottom sediments. According to the sedimentary structures (anti-ripplets, rill marks, desiccation cracks), the water was shallow and subject to frequent drying-out. The fish died in the residual pools of water lacking in oxygen because of a partial emergence of the environment. Anaerobic conditions in the muds favoured their preservation.

Eventually, compaction of the sediment flattened the fossils.

4. The Environment

The Achanarras beds probably extend laterally for more than 100 km. No marine fossil has ever been found in these horizons, which are the deposits of a **great lake**, Lake Orcadie, whose waters were shallow and underwent local periods of drying-out.

The sedimentation sometimes shows a rhythmic nature. Possibly an algal bloom developed at intervals. This intense organic activity raised the pH and caused the precipitation of carbonates. In dying, the algae raised the organic content of the bottom muds. This phenomenon is well known in shallow water lakes in southern Australia, possibly being repeated annually. This would explain the origin of the laminae which are formed by an alternation of carbonate and organogenic units. The quartz grains, which form a considerable part of the rock, doubtless represent the load of rivers which flowed into the lake.

The Orcadian Lake sheltered a varied fish fauna. The commonest forms are the dipneusts whose requirements were close to those provided by the environment. The twofold method of respiration of these fish allowed them to survive during the periods when the waters were restricted or dried up.

To a certain extent, the other fish must also have been able to tolerate a temporary reduction in dissolved oxygen. Such an adaptation heralds the conquest of the land by the vertebrates.

III. The Rhynie Peat Bogs

The Rhynie horizon, west of Aberdeen in Scotland, was discovered in 1913. It is part of a huge outcrop of Old Red Sandstone attributed to the Lower Devonian.

1. The Sediment

The Rhynie cherts overlie a thick series of red comglomerates, sand-stones and shales. They consist of beds formed by an accumulation of **silicified plants** separated by sandy horizons which are also silicified. The fossiliferous sediment is grey-blue in colour and semi-translucent. When wet, it reveals morphological details of an astounding delicacy (Pl. I, Fig. 4).

2. The Fossils

a) The Fauna

The fauna contains aquatic animals only.

— *Lepidocaris* is a **branchiopod crustacean** barely 3 mm long as an adult (Fig. 103A). It moves in water by means of its long biramous antennae. There are both male and female forms. Stages in development from a juvenile, which has only 4 body segments, to the adult, which has 20, have been found.
— Several species of **spider**, belonging to the trigonotarbids, the araignids and the acarians, occur.
— *Crania rhyniensis* has been compared to the **eurypterid** merostomes.
— The Rhynie horizon also contains the oldest **insects** known, in particular an aquatic collembole, *Rhyniella*.

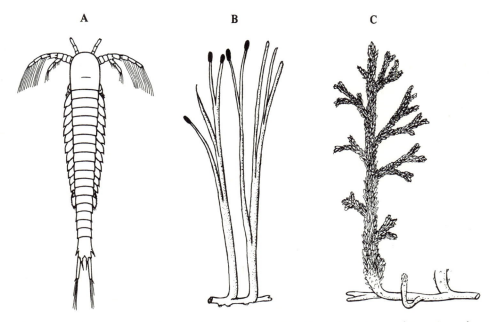

Fig. 103A—C. Fauna and flora of the Devonian at Rhynie. **A** *Lepidocaris* (size: 3 mm). (Scourfield 1926). **B** *Rhynia;* **C** *Asteroloxon*

b) The Flora

The Rhynie plants are abundant and remarkably well preserved. They belong to the **psilophytid** group and are amongst the oldest vascular plants known (Fig. 103B,C).

— *Rhynia* had dichotomous, cylindrical, erect, aerial stems some decimetres long. They are leafless and develop from underground rhizomes.
They have stomata in their cuticle and terminal sporangia. Minute anatomical details (shape of the cells, structure of the vessels, etc.) can be
studied in thin sections. **Cankers** have been described on some stems.

— *Horneophyton,* which looks rather like *Rhynia,* must have formed
very dense thickets since sometimes it is the only constituent of the
sediment. Its tuberous rhizomes frequently shelter an intercellular
fungus, *Palaeomyces.*

— *Asteroloxon* is a more complexly organised psilophyte than the two
preceding forms. The rhizome produces branching aerial stems which
carry small leaves. The reproductive organs are leafy spore-bearing
axils.

In the Rhynie flora, *Rhynia* and *Horneophyton* clearly correspond to
very early forms, with a simple anatomy. They co-exist with more highly
developed plants such as *Asteroloxon.*

Aquatic algae, associated with other plants, occur at several horizons.

3. Fossilisation

The **silicification** which is responsible for the beautiful preservation of
the fossils must have begun soon after the deposition of the plant debris,
since it stopped decomposition of the tissues by bacteria and saprophytic
fungae.

The silica-rich waters which impregnated the peat-bog several times
came from hot springs whose existence was no doubt associated with the
intense volcanic activity indicated by the andesites interbedded in the
Old Red Sandstone.

4. The Environment

The vascular plants, the aquatic fauna and the algae must have belonged
to neighbouring biotopes. The first represent the dense **peat-bog** vegetation. Toxic gases from fumaroles are probably the cause of the cankers
developed on the aerial shoots. The climate must have had a prolonged
dry season, since the rhizomes of *Horneophyton* are tuberised.

The crustacea, the algae, the spiders and the insects lived in nearby
ponds. From time to time, hot springs associated with acid volcanism
discharged silica-rich water and flooded the neighbouring peat-bogs. The
fauna and flora were thus rapidly silicified.

The sand forming the sandstone layers was doubtless blown into the peat-bogs by the **wind** and the same process would disseminate the eggs of *Lepidocaris,* as it does for the present-day branchiopods. This maintained the population of the pools of water and accounts for the low diversity of animal life in Devonian ponds and lakes.

IV. Conclusions on the Environments of the Old Red Sandstone Continent

Throughout Devonian times, the erosion of the Caledonian chain fed the huge regions of the North Atlantic continent. A complex of fluviatile channels with wandering courses laid down a huge mass of red sediments. Beside the zones of active sedimentation were temporary accumulations of water, lakes, ponds and swamps, populated by crustaceans, spiders, merostomes, insects, agnathans and fish. Vascular plants thrived close by. The hot climate, with its contrasting seasons and long periods of dryness, reddened the sediments.

It was in such an environment that the flora and fauna took their first steps towards the conquest of the land.

Chapter 9 The Decazeville Coal Basin

The Decazeville coal basin lies in a depression in the basement rocks of the southwest part of the Massif Central in France. This depression is elongated north-south and is about 20 km long by 8 km wide. To the west, it is bounded by the major dislocation of the Grand Sillon Houiller. During mid and late Stephanian times, detrital sediments with frequent intercalations of coal were deposited in this basin. The sediments can be divided into six supergroups, with a total thickness of 1800 m (Fig. 104).

I. The Sediments

Terrigenous and plant debris filled the pre-Stephanian depression.

1. Petrology

a) Terrigenous Sediments

The terrigenous sediments consist of conglomerates, sandstones and siltstones.

The constituents of the conglomerates have been derived from the crystalline rocks and volcanics at the edge of the basin. Sometimes, they are **monogenetic breccias** which formed as scree slopes. More often, the pebbles in the conglomerates are well-rounded. They vary in size up to several tens of centimetres and are typically fluviatile. There is also a whole range of sediments from coarse sandstones to fine shales: these latter contain carbonaceous fragments and, occasionally, bitumen.

Two clay-grade rocks are worth noting:
— **Tonsteins** formed of kaolinite derived from volcanic ashes; because of their great lateral extent, they are excellent stratigraphic markers;
— red **bauxitic clays** which formed as a result of a lateritic alteration of the regions bordering the basin.

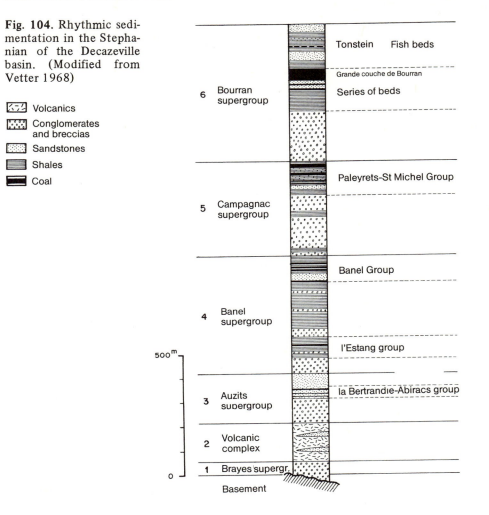

Fig. 104. Rhythmic sedimentation in the Stephanian of the Decazeville basin. (Modified from Vetter 1968)

Volcanics

Conglomerates and breccias

Sandstones

Shales

Coal

Tonstein   Fish beds

Grande couche de Bourran

6 Bourran supergroup   Series of beds

5 Campagnac supergroup   Paleyrets-St Michel Group

4 Banel supergroup   Banel Group

                               l'Estang group

                               la Bertrandie-Abiracs group

3 Auzits supergroup

2 Volcanic complex

1 Brayes supergr.

Basement

500^m

0

b) Plant-Derived Sediments: Coals

Under the microscope, it is easy to see the vegetable origin of coal. The main constituents are cuticle, spores, fungal sclerotes and chunks of wood. **Pyrite** grains are disseminated through the sediment.

The nature and relative abundance of these constituents controls the various types of coal:

— **vitrain** is shiny because of an abundant amorphous cement formed from the decay of plant tissues;

— **fusain**, being rich in wood debris, is friable and fibrous;

— **durain** is matt and has a grainy fracture; it is composed mainly of spores and cuticle;

– **clarain**, which is rare in the Decazeville coals, has a heterogeneous appearance because of the variety of fragments within its cement.

The Decazeville coal is a soft coal with a low calorific potential. It contains 29%–31% volatiles.

Concretions and lenses of iron carbonate (siderite) occur at all levels in the formation. Limestones and dolomites are rare.

2. Stratinomy

a) Bedding

The terrigenous deposits and coal seams are lenticular (Fig. 105). The beds vary considerably in thickness from east to west; their lateral extent is greatest along the line of the basin.

The breccias and conglomerates are frequently more than 100 m thick and fill channels cut in the older rocks. In most cases, the sandstones form beds a few centimetres to some metres thick; cross-bedding is general. Bedding in the shales is more nearly horizontal and picked out by plant debris. **Roots in their position of life** are common and mark the position of fossil soils.

The coal seams vary in thickness from a few centimetres to several tens of metres. The Grande Bourran seam is exceptional in reaching 40 m. The upper surface or roof of the coal seams is, in general, well-defined, but its lower surface or floor is more difficult to identify, since lenses of shale of varying carbon content form a passage from barren sediment to coal. The coal itself has a fine lamination due to the accumulation of plant debris. The coal seams and their associated barren beds are the **groups** sought in exploration.

The many workings in the Decazeville basin have proved the frequency of lateral facies changes. The thickness of the coals is greatest in the centre of the basin; to east and west, they finger out or pass gradually into barren beds (Fig. 105). While the Capagnac group was being deposited, for example, the width of the elongated basin in which the plant debris was accumulating was barely 1500 m.

b) Rhythmic Deposition

The infill of the Decazeville basin consists of six supergroups which correspond to former sedimentary events (Fig. 104). After an early conglomeratic deposit, which rests directly on crystalline basement, there was a series of acid volcanic extrusions. The coal seams occur only in the four

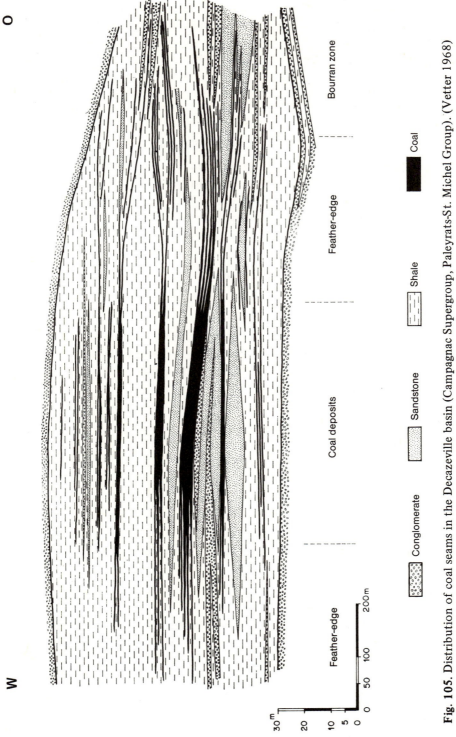

Fig. 105. Distribution of coal seams in the Decazeville basin (Campagnac Supergroup, Paleyrats-St. Michel Group). (Vetter 1968)

following units (Auzits Supergroup, Banel Supergroup, Campagnac Supergroup and Bourran Supergroup) and become thicker towards the top of the series.

From bottom to top, each of the four supergroups contains a similar succession of deposits:
- a **coarse detrital** series, which is often conglomeratic and represents fluviatile deposits;
- a **sand/shale** complex;
- a **coal**-rich unit with many intercalations of shale.

This sequence of lithological units indicates a gradual weakening of the energy of the sedimentary basin and the establishment of conditions favourable to the deposition of plant horizons. The frequency of seat-earths proves that this took place in shallow water.

II. The Fossils

1. The Flora

Coal deposits are characteristically very rich in plant debris. One hundred and fifty four species have been collected in the Decazeville basin; they come mainly from the shales which form the roofs of the coal seams and are usually found as small fragments.

a) Thallophytes

Thallophytes are represented by **fungae** parasitic on the fronds of ferns and the leaves of cordaitales.

b) Bryophytes

Some plant moulds have been attributed to the liverworts.

c) Pteridophytes

Lycopodiales are rare in the Decazeville basin. Those present are arborescent forms *(Lepidodendron, Sigillaria)* known from fragments of their trunks, their fruiting cones and their roots (Fig. 98).

Equisetales are abundant, with very tall species *(Calamites)* living next to herbaceous plants *(Sphenophyllum)*.

Filicales, which are often arborescent, are represented by numerous species of the genus *Pecopteris*.

d) The Prephanerogames

The pteridosperms combined a fern-like leaf with reproduction "by seeds". These "seed ferns" include many genera (*Alethopteris, Neuropteris, Odontopteris, Dicksonites,* etc.).

The **cordaitales** had tall, thin trunks with long, simple leaves and very large flowers (Fig. 98).

e) The Phanerogames

The phanerogames are represented only by a few conifers of the genus *Lebachia.*

Many isolated "**seeds**" have been assigned to the prephanerogames, as have **roots** standing more or less vertical to the bedding in the shales and sandstones. These are attributed to various pteridophytes. A great variety of spores has also been extracted from the shales and the coals.

2. The Fauna

Compared to the diversity of the plants, the fauna of the Decazeville basin is very impoverished. Apart from insect wings, it all comes from the Bourran Supergroup.

a) Terrestrial Animals

The terrestrial animals include a **spider** (*Trigonotarbus*) and various **insects** belonging to the palaeodictyoptera and the cockroaches.

b) Aquatic Animals

The **bivalves** are represented by the genus *Anthracomya,* a burrowing form sometimes found associated with its burrow. Some freshwater **ostracodes** have also been reported. The main part of the aquatic fauna, however, consists of freshwater **fish** (acanthodians, elasmobranchs, palaeoniscids).

III. The Environment

1. The Limnic Nature of the Sedimentary Basin

Because of the coal workings, the lateral extent of the Stephanian sedimentation in the Decazeville basin is now well known. It is shaped like a large north-south gutter. Its coarse detrital deposits were carried in from the surrounding crystalline uplands by a complex of watercourses [1].

In contrast, the deposition of the fine sediments of the shales and the coal seams took place in quiet water. The flora and aquatic fauna suggest that the water was **fresh**. It was generally shallow, probably not more than a few decimetres, as indicated by the frequency of plant roots in situ.

Apart from the coal seams, the presence of pyrite and the preservation of carbonaceous material suggests that conditions were anaerobic. This would explain the rarity of benthic organisms. These conditions, hostile to the development of a fauna but favourable to the preservation of plant material, must have been maintained for a considerable time to account for the thickness of the coal seams.

The various plants, whose remains produced the coal, must have lived close to the sedimentary basin. Their fragmentation is as much due to their being soaked in water as to mechanical breakdown during transport. The shales with roots indicate that a swamp flora was present. One might describe the vegetation as **subautochthonous**.

These observations suggest that the Decazeville basin acted as a **collecting area** where watercourses dumped the load they had carried from the surrounding uplands, which were covered by abundant vegetation. Swamps developed in the low-lying parts during quiet periods and a great thickness of plant remains was able to accumulate.

2. Sedimentary Processes

The frequency of **seat earths** interbedded with the coal seams suggests that the body of water remained shallow during the whole period of formation of the coal deposits. In these conditions, the thick accumulation of sediments can only be explained by subsidence of the depositional region. Faults continually rejuvenated the uplands bordering the basin and intense volcanic activity took place, maintaining the subsidence throughout the Stephanian. The rhythmic sedimentation which can be seen in the various supergroups means that the **subsidence** had to **vary in speed**.

---

1 Fayol has reconstructed the exact topography of a lacustrine region fed by watercourses in the Commentry coal basin, in the north of the Massif Central

Periods of rapid subsidence produced a need for sediment: high capacity watercourses flowed from the neighbouring uplands and turned the area into a vast collecting basin. These sediments are now the conglomerates at the base of the supergroups. When subsidence slowed, the sediments became finer and passed from sandstones into shales. Finally the low-lying swamps developed. The subsidence of the basin was thus cancelled out for long periods of time by the plant deposits which gave rise to the coal. As one went from the centre of the basin and approached the source areas, large tracts of sterile land became more and more common.
of erosion, a new sedimentary cycle began (Fig. 78).

3. The Climate

Lateritisation contemporaneous with coal formation suggests a **tropical climate**. High temperature and humidity produced conditions favourable to the spread of plants and to intense geochemical alteration. The iron freed by this alteration was partly concentrated in a skin which built up beneath the forests bordering the basin, while the iron in solution accumulated in the basin itself as siderite.

4. Conclusion

The Decazeville coal basin is the remains of an intramontagne depression of limited extent, hollowed out in a zone of low resistance in the crystalline basement. During Stephanian times, **active tectonism** initiated the accumulation of a thick series of detrital and plant sediments derived from the nearby uplands. The periodic subsidence of the area gave rise to rhythmic sedimentation. At times of low tectonism, the remains of a varied plant community, whose growth was favoured by a tropical climate, soaked in swampy, badly oxygenated hollows. With time, the activity of anaerobic bacteria followed by diagenetic processes enriched the carbon content of the sediments and transformed them into coal.

Chapter 10   **The Grès à Voltzia Delta**

The Grès à Voltzia correlates with the upper part of the **Bunter Sandstone** in the east of France. It covers the passage from the continental formations of the Triassic red sandstones to the marine sedimentation of the Muschelkalk. On average, it is about 20 m thick and actually consists of two very distinct units: the *Grès à meules* (millstone grit) overlain by the *Grès argileux* (argillaceous grit). From its palaeontological and sedimentological content, the Grès argileux marks the beginning of the Muschelkalk Sea transgression, while the Grès à meules represents the last stage of the Bunter fluviatiles. Only this latter formation will be considered in this chapter.

In the north of the Vosges, between Saverne and Sarre-Union, the Grès à Voltzia is still worked in several quarries to make millstones and for construction (Adamswiller, Bust, Hangviller, Petersbach, etc.). There the Grès à meules is about 12 m thick.

I. The Sediment

1. Petrology

a) The Sandstones

The Grès à Voltzia is fine-grained (mean grain size Q_2 ranges between 0.10 and 0.20 mm), well-sorted, often pink or red but sometimes grey. It contains 20% to 30% potash feldspar. The grains are poorly rounded and the sparse pelitic matrix fixes the ferric oxides which give the rock its colour.

These features are characteristic of a **juvenile sediment**, submature or mature in Folk's classification (Fig. 52). Controlled alteration of the original material has retained the alkali feldspars and it is not possible to suggest that a tropical climate existed in the source areas. Moreover, the energy available during transport was not enough to entirely eliminate the clay fraction or round the grains.

b) The Shales

The shales of the Grès à Voltzia consist mainly of illite and are green or red. Some green horizons are rich in organic matter and pyrite. There is a whole range of lithologies from shale with virtually no sand to argillaceous sandstones.

c) The Carbonates

The carbonate horizons, which are now usually breccias, were originally either sandy dolomites or carbonate-enriched sandstones where calcite and dolomite were present in equal parts.

2. Stratinomy

a) Bedding

Bedding is lenticular throughout the Grès à meules. The lateral extent of the beds of sandstone or clay, as well as that of the carbonate horizons, varies from a few metres to more than a hundred metres.

The sandstone lenses can be several metres thick and cut down into the underlying deposits. There are two types of sandstone:
- the **plant-bearing sandstones**, coarse, with mud-pebbles and vertebrate debris; bedding is indistinct;
- the **clean-washed sandstones**, actively worked because they are fine-grained; pebbles and fossils are absent; they are very finely horizontally or cross-bedded.

The shale horizons are several centimetres or decimetres thick. They are often horizontally laminated, on a millimetre scale; each lamina is vertically graded, its base is high in quartz and feldspar, while clays and organic matter are concentrated near the top.

b) Sedimentary Structures

The lower, erosive surface of the sandstones has a rich assemblage of sedimentary structures (Fig. 106): flute casts and groove casts. The upper surface of the lenses is covered in ripple-marks (Pl. II, Fig. 2). The sandy laminae show parting lamination and crescent-shaped scour marks. The succession of structures within a bed indicates a decrease in the strength of the depositing current with time. Statistical analysis shows that these

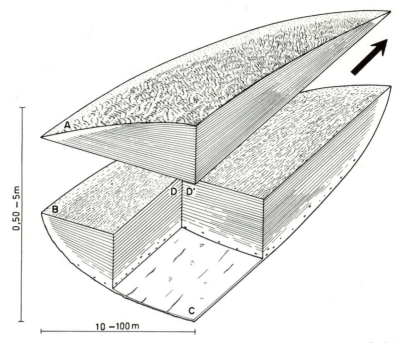

0,50 – 5m

10 –100 m

Fig. 106 A—D. Block diagram of a fluviatile sand body in the Grès à Voltzia of the Vosges (the current decreased from the bottom to the top of the bed). **A** upper surface with ripple marks; **B** lamina surface with parting lineation; **C** lower surface with current structures; **D, D'** section through the laminae. *Arrow* indicates the current direction. (Gall 1971)

currents flowed from west to east, from uplands on the site of the present Paris Basin (gallic province) towards the German sea.

Flaser bedding occurs when the sediment is rich in clay. Sets of laminae are sometimes affected by sub-aquatic slumping. Desiccation cracks, or, more rarely, salt pseudomorphs, occur on the tops of the clay lenses.

3. Geochemistry

The trace element content of the clay minerals indicates that the clay and carbonate horizons were deposited in higher salinity environments than the sandstones. The **boron** content of the pelitic deposits is particularly high (300 to 400 g/t).

II. The Fossils

The Grès à Voltzia is one of the most fossiliferous horizons in the germanotype Trias, as much for the diversity of organisms present as for their
preservation. Its palaeontological treasures have largely been described
by the patient research of Grauvogel.

1. The Fauna

a) Foraminifera

The rare foraminifera are represented by forms with an agglutinating or
calcareous test (ammodiscids, lagenids). They are marine.

b) Coelenterates

Medusids belonging to the order of limnomedusids *(Progonionemus)* have
been described (Fig. 107). On some specimens, the rings of stinging cells
are still visible on the tentacles. Very large individuals show four gonads
spirally around the radial canals.

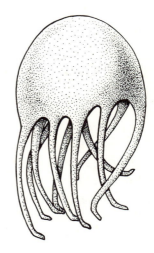

Fig. 107. Juvenile medusid *(Progonionemus)* from
the Grès à Voltzia of the Vosges

c) Brachiopods

The brachiopods are represented only by **lingulids,** often preserved in
their life position. This genus is characteristic of shoreline waters with
fluctuating salinity.

d) Annelids

Body fossils of annelids are of the free-living type (*Eunicites,* aphroditids). Fixed forms *(Spirorbis)* were attached to plants or to the shells of bivalves.

e) Molluscs

The shells of the bivalves are thin and their size generally smaller than that of the same species coming from fully marine waters *(Myophoria).* Some were burrowers *(Homomya).* Gastropods are rare *(Naticopsis).*

f) Arthropods

Arthropods form the most abundant group in the fossiliferous horizons of the Grès à Voltzia. The chitinous shell of these animals is admirably suited to fossilisation. **Moults** have frequently been preserved.

i) Chelicerates

The aquatic forms, the **limulids,** are especially common and even their locomotion traces are preserved. **Scorpions** dominate the terrestrial forms (Pl. II, Fig. 3). Spiders are rarer.

ii) Crustacea

Triops cancriformis is a branchiopod crustacean identical in every way to the present species which lives in temporary pools of water. It is an original "living" fossil, which first evolved at the beginning of the Mesozoic.

The bivalved carapaces of **estheriids** (Fig. 28) are the most frequent fossils in the Grès à Voltzia. They are conchostracean crustacea several millimetres long.

Ostracods are abundant only at certain specific horizons.

The higher crustacea include mysidacids (prawns), isopods and decapods (*Antrimpos* among the swimmers, *Clytiopsis* among the walkers) (Fig. 108).

Other crustacea, of more uncertain affinities, are unique to the Grès à Voltzia *(Euthycarcinus).*

Fig. 108. Decapod crustacean *(Antrim-pos)* from the Grès à Voltzia of the Vosges. (Gall 1971) (× 2.5)

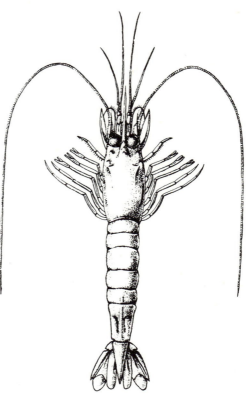

iii) Myriapods

Some diplopod myriapods have been found in the Grès à Voltzia. They were vegetarian.

iv) Insects

Insects are very abundant. They are mainly known from their wings, but also from their aquatic larvae and their eggs. They belonged to various groups: the emphemerids (mayflies), the odonatids (dragonflies), the blattarids (cockroaches), the coleopterids (beetles), the dipterids (flies), the hemipterids (bugs), etc.

g) Fish

Chrondrosteans *(Dipteronotus),* holosteans and coelacanths are among the fish represented; juveniles are especially abundant.

h) Tetrapod Vertebrates

The tetrapod vertebrates are mainly represented by the remains of amphibians (stegocephalids) and rare reptile fragments.

2. The Flora

The flora of the Grès à Voltzia is abundant but of low diversity. Arborescent forms are rare and the plants are mainly bushes. Some plants have **xeromorphic characteristics**.

a) Pteridophytes

The pteridophytes are represented by equisetales *(Schizoneura, Equisetites)* and ferns *(Anomopteris);* their rhizomes are sometimes preserved in situ in the sediment.

b) Prephanerogames

Cycads and ginkgos have been described.

c) Phanerogames

The characteristic flora of the Grès à Voltzia is made up of conifers including *Voltzia heterophylla* (Fig. 109); they co-exist with other gymnosperms *(Aethophyllum).* Male and female reproductive organs, together with **seeds**, are frequently associated with them. **Pollen** can be studied in situ in the cones.

Phanerogam **root systems** fossilised in position of life prove that at least part of the flora is autochthonous.

3. Evidence of Biological Activity

a) Eggs and Clutches

One of the most remarkable types of fossil from the Grès à Voltzia is the **eggs** attributed to **insects** (Figs. 29, 110). The eggs measure about 0.25 mm in diameter and are protected by a chitinous shell which opens along a median slit on hatching. They were stuck together by a sheath of mucilage so that each clutch looks like a necklace or club, reminiscent of the eggs of some present-day chironomids. The eggs, between 500 and 3000 per clutch, are usually unhatched, and their internal structures, including embryos, have been described.

Eggs are also found within the valves of **estheriids** (Fig. 28). Numerous small eggs can be interpreted as eggs intended for immediate development to ensure the continuity of the species within the same environment, while the scarcer, larger eggs have a longer life, being distributed by the wind when the environment dried out. In this way, the population of temporary pools of water was protected.

b) Coprolites

Lumps rich in crustacean debris have been interpreted as fish or amphibian coprolites.

c) Trace Fossils

The trace fossils include crustacean dwelling burrows *(Rhizocorallium),* arthropod trails (limulids), reptile tracks *(Cheirotherium)* and rare feeding systems *(Planolites).*

To summarise: in the fossiliferous horizons of the Grès à meules, **terrestrial organisms** (plants, scorpions, spiders, myriapods, insects, amphibians, reptiles) lived alongside **aquatic organisms** (foraminifera, coelenterates, brachipods, annelids, molluscs, limulids, crustaceans, fish). Amongst the latter, *Lingula* is characteristic of water with a fluctuating salinity, while the foraminifera and some of the molluscs indicate a marine environment. Both the flora and fauna are rich in individuals but poor in species. What is more, many forms in the aquatic fauna are of rather small size. They might either be dwarfs, or individuals which died before they reached their adult size. All these observations suggest that living conditions were very difficult in the Grès à Voltzia environment.

III. The Environment

Both in the variety of its lithologies and in the distribution of its organisms, the Grès à Voltzia is a patchwork of environments. These can be grouped into three facies: the deposits of fluviatile channels, the sediments of temporary pools and the deposits of littoral mudflats.

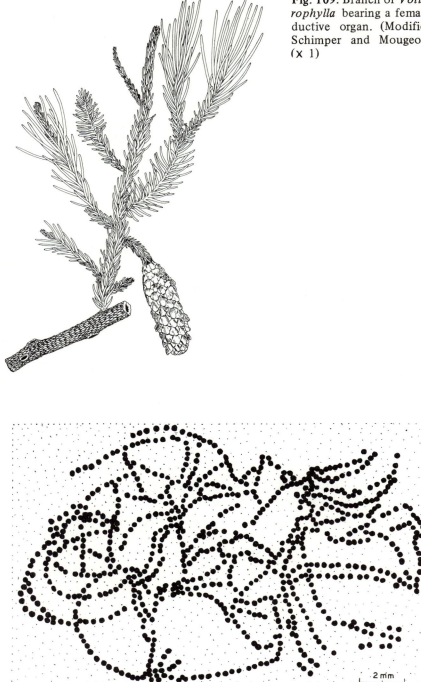

Fig. 109. Branch of *Voltzia hete-rophylla* bearing a female repro-ductive organ. (Modified from Schimper and Mougeot 1844) (x 1)

Fig. 110. Clutches of insect eggs *(Monilipartus)* from the Grès à Voltzia of the Vosges. (Diameter of each egg 0.25 mm)

1. Fluviatile Channels

The sandstone lenses with an erosive base were deposited in channels. Within each bed, the decrease in the strength of the current through time is recorded more as a succession of sedimentary structures than as a decrease in grain size (Fig. 106). Palaeocurrent studies show that flow was from the west towards the east. Moreover, the spread of measurements and the fine grain size of the transported material indicate that the river complex wandered and formed meanders.

The poorly sorted deposits of the plant-bearing sandstones are the result of a **flood from a watercourse**. In its passage, this ripped up blocks of mud from its banks and moulded them into pebbles. It also dragged in plants and amphibians living on dry land. The whole sedimentary load was dropped suddenly, further downstream, without any appreciable sorting, so that the plant remains lie oblique to the bedding of the sandstone.

The clean sandstones, on the other hand, are well sorted and contain neither mud pebbles nor fossils. They form large lenses where the material has been winnowed, abraded and sorted over a long period of time in a high energy environment, probably at the **channel mouths**.

The continental origin of the water is confirmed by the presence of amphibians and the low boron content (100–200 g/t).

2. Temporary Pools of Water

The lenses of shale were deposited in bodies of water of small lateral extent: basins or abandoned channels. The main part of the flora and fauna of the Grès à Voltzia comes from here (Fig. 97).

a) Water Salinity

The high boron levels in the clay minerals (300–400 g/t) and the sporadic occurrence of **salt pseudomorphs** suggest that the water was of a higher salinity than in the channels. Moreover, trace element concentrations show that the salinity fluctuated within a single shale lens.

Amongst the aquatic fauna, the absence of stenohaline organisms (brachipods, cephalopods, echinoderms) indicates that conditions were not fully marine. On the contrary, the presence of euryhaline forms such as lingulids, the dwarfing which seems to have affected many genera, and the low diversity in contrast to the high faunal density are characteristic of **brackish water** populations.

b) The Ephemeral Nature of the Pools

The shallowness of the water is supported by a whole series of observations: the frequency of oscillation ripples and desiccation cracks, the presence of plants, terrestrial animals and lingulids, reptile footprints, clutches of insect eggs, etc. The tops of the shale lenses frequently also have **plant roots in life position** or desiccation cracks. Moreover, as one goes upwards in these horizons, a diverse aquatic fauna is gradually replaced by a swamp vegetation, while trace element content, especially boron, increases. This indicates that, with time, the pools became isolated, the salinity rose and they dried out, the end of their life being marked by colonisation by plants. In some cases, the solubility limits of carbonate and halite were reached, and they precipitated out.

The abundance of **estheriids** in the shale horizons is significant. The life cycle of these crustaceans only lasts a few weeks and the eggs which were stuck to their valves were rapidly blown away and spread by the wind as the environment dried out.

c) The Hydrodynamic System

The clays were deposited in quiet water. The grading within many of the laminae can be attributed to periodic influxes of detritus. However, apart from a few turbulent episodes, water movement remained low during the deposition of the shale horizons, which partly explains the excellent state of preservation and absence of orientation of the fossils.

d) Oxygenation

The conditions in the muds at the bottom of the pools were often **reducing** and unsuitable for lasting colonisation by an endofauna, as suggested by the preservation of the fine bedding and the sporadic presence of pyrite.

The water itself, however, was sufficiently aerated to allow the spread of a rich fauna of vagile benthos and fish. This oxygenation occurred by means of periodic influxes of water and possibly also by the action of floating aquatic plants.

Equilibrium was rarely achieved between water supply and evaporation in these shallow pools. Over the years, there were great variations in temperature, oxygenation and salinity. These fluctuations in physico-chemical conditions rebounded on the aquatic populations and caused **seasonal mass mortality**. All these phenomena are characteristic of brackish or hypersaline environments at the margin between land and sea.

3. Littoral Muds

Marine influences are more obvious in some horizons of the Grès à Voltzia, especially in the sandy dolomites or dolomitic sands which contain foraminifera and molluscs, and the shale lenses with a fauna of lingulids and burrowing bivalves, than they are in others.

Such deposits were formed close to the shore, in littoral muds where the oxygenation of the water was more favourable to an endofauna than in the temporary pools.

IV. Fossilisation

1. The Plant-Bearing Sandstones

The palaeontological content of the plant-bearing sandstones consists of plant debris (equisitales, ferns, etc.) and amphibian bones (stegocephalids); they are disarticulated and broken. Plants and animals had been dragged in from the eroded banks of the watercourses and underwent prolonged transport before being dumped by a flood. They accumulated in the channels as **thanatocoenoses.**

2. The Clay Horizons

Compaction has flattened all the fossils in the clay horizons.

a) Palaeobiocoenoses

The aquatic fauna in the clay horizons shows many signs of being **autochthonous.** In the first place, the preservation of many structures as fragile as the bell of a medusid or crustacean appendages suggests that the organisms have not undergone any appreciable transport. Bivalves with their shells still articulated, orientated concave-upwards, confirm this conclusion.

The constant association in the same horizon of clutches of eggs, different larval stages, adults and moults of arthropods (limulids, crustacea, insects) proves that these animals lived where they were fossilised. This is supported by the presence of lingulids in their life position and tracks of limulids and crustacea.

The equally remarkable state of preservation of the terrestrial fauna (scorpions, spiders, myriapods, insects) and plants with their reproductive organs indicates that they lived on the banks or on islands extremely close to the water into which they accidentally fell. Rhizomes and root systems in situ in the sediment support this interpretation.

The aquatic fauna of the shale horizons, and, to a lesser extent, the plants and terrestrial organisms represent true **palaeobiocoenoses**. Actually, detailed study of the fossiliferous horizons indicates two types of palaeo-biocoenosis: one where crustacea are dominant and one characterised by lingulids and bivalves which are typical of the more brackish waters of the littoral mudflats.

b) Mass Mortality of the Aquatic Fauna

The aquatic organisms normally occur at the top of the graded laminae. They include individuals at different stages in their ontogeny, often concentrated in groups. Within the egg clutches of insects, most of the eggs are unhatched. These features suggest a fauna which has been suddenly exterminated.

The **laminae** themselves provide an explanation of this phenomenon. Each sedimentary unit, whether a few millimetres or a few centimetres thick, shows vertical grading and thus corresponds to a single sedimentary influx. The arrival of the water which laid down this deposit also brought in oxygen and nutrients and aquatic animals could therefore proliferate. Then the influx slowed, the clay particles dropped out and the water became stagnant. The water warmed up and lost oxygen. Sometimes hydrogen sulphide was liberated or the environment partially dried out. This caused mass mortality amongst the aquatic fauna, which accumulated at the top of the lamina. With the next influx, the cycle began again.

The **intermittent supply of detrital material** which controlled the succession of laminae was itself probably controlled by flooding of the river complex, by the tides or by rainfall. Clearly the clay lenses which lack graded laminae are commonly poor in aquatic animals, the corresponding pools of water being shielded from the influence of intermittent supplies of sediment.

c) Preservation of Organic Matter

The fossilisation of the gelatinous bodies of medusids and the clutches of insect eggs can only be explained by desiccation in air. Under these conditions the organic matter was mummified and took on the consistency of parchment.

Arthropod chitin and plant material found a favourable environment for preservation in the reducing conditions within the fine-grained sediment.

These conditions are indicated by the rarity of endobiontic organisms and by the presence of pyrite. On the other hand, such a sediment was hardly good for the preservation of calcareous shells, and one finds the valves of bivalves reduced to their skin of conchiolin.

The burial of the animals was rapid. Within a 60 cm thick lense, it is possible to follow the biological cycle of the gymnosperms which grew nearby (Fig. 97). In the lower part of the unit, pollen grains corresponding to the flowering season are abundant, then come scattered scales from the male cones and finally seeds. Such a succession could have taken place between the spring and the autumn of a single year. This indicates that the rate of sedimentation was high and that one lamina was deposited in a few weeks.

To summarise: the fossilisation of the fauna and flora of the Grès à Voltzia took place in particularly favourable circumstances: **rapid burial in a fine-grained sediment and a reducing environment**. This explains the astonishing chemical stability of the saturated hydrocarbons extracted from some equisitids.

V. Conclusion

The different habitats of the Grès à meules can easily be placed in geographical relation to one another. Fluviatile channels wandered on a flat alluvial plain forming meanders. Between the watercourses, in abandoned channels or in depressions on the flood plain, pools of water developed, became concentrated and dried out at varying speeds. They were colonised by euryhaline organisms from the nearby sea and were bordered by a swamp flora. Littoral mudflats formed further downstream.

The interfingering of these facies is typical of the subaerial part of a **deltaic environment** (Fig. 111). The bar facies of the clean sandstones formed the beginning of the delta front sedimentation, while further to the east, in the Black Forest and the Odenwald, the red shales of the Röt correspond to the prodelta muds. One can easily imagine that this changing semi-land area was frequently invaded by the sea.

Pole positions deduced from palaeomagnetic studies of the rocks, the red colour of the sediments and the xeromorphic nature of the flora indicate that during Grès à Voltzia times the **climate** was hot, with alternate dry and wet seasons. These gave rise in their turn to the transport of detrital material, the filling of channels and pools and the spread of life, to the isolation of the ponds, their drying-out and the death of the aquatic fauna.

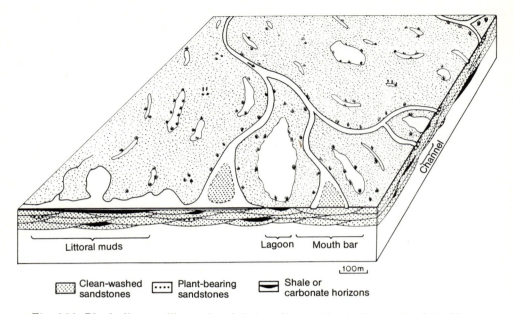

Fig. 111. Block diagram illustrating deltaic sedimentation in the north of the Vosges during early Grès à Voltzia times. (Gall 1971)

At the top of the Grès à meules is a horizon of plant roots in life position. It marks the decrease in detrital influx and the establishment of vast swamps: the delta was dead. Sedimentation could not keep pace with regional subsidence and the delta flat subsided steadily into the sea: it was the beginning of the transgression of the Muschelkalk sea.

Chapter 11   **The Reefs of Hoher Göll**

The Hoher Göll massif, which lies about 20 km south of Salzburg in Austria, is a part of the Berchtesgaden Alps, which themselves form part of the North Calcareous Alps.

In Triassic times, it formed part of a vast epicontinental platform which bordered Tethys and was the site of thick carbonate sedimentation (limestones and dolomites). In late Triassic times (Carnian and especially Norian and Rhaetian) there was a period of reef formation which produced a series of very diverse facies whose significance is now well understood (Fig. 112).

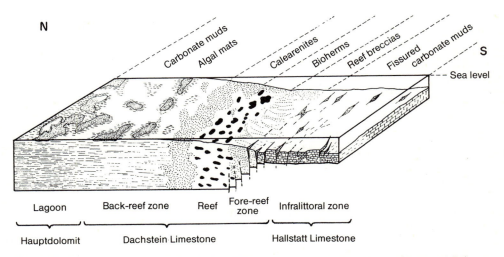

Fig. 112. Palaeogeography of the Hoher Göll region in late Norian times. (Zankl 1971)

I. Hallstatt Limestones

1. The Sediment

The Hallstatt limestones are made of **micrite**. The insoluble residues, which form a relatively low proportion of the sediment, consist of clay minerals, quartz and iron oxide.

Bedding is thick, planar and parallel. Local calcite solution during diagenesis has given rise to calcareous nodules. Hardgrounds and intraformational breccias occur several times within the series.

2. The Fossils

The micritic matrix of the rock is rich in **micro-organisms**: planktonic forms (foraminifera, radiolaria), ostracodes, and mollusc and echinoderm debris. Conodonts are concentrated in the insoluble fraction.

The **macrofauna** accumulated in contemporaneous **fissures** in the sediment, which acted as important traps for the shells of cephalopods and bivalves. These latter also occur in thick banks *(Halobia, Monotis)* where the valves frequently lie concave upwards. Doubtless they lived a pseudo-planktonic life. Ammonite shells *(Arcestes)* are sometimes found standing vertical in the sediment. These features are typical of sedimentation in quiet water.

Encrusting foraminifera and thallophyte borings on hardgrounds indicate that the water was well oxygenated at depth.

3. The Environment

The micritic nature of the sediment, the regularity of the bedding, the pelagic fauna and the orientation of the mollusc shells indicate that the limestones were deposited in a **quiet sea**. The interbedding of the Hallstatt Limestones with the Dachstein reef facies indicates that the environment lay on the outer part of the continental shelf, beyond wave base, where the water depth was considerable (50–200 m) while still remaining aerated. The intraformational breccias, formed by moderate tectonic activity, indicate that the sediment was rapidly lithified.

The Hallstatt Limestones are overlain by the marly beds of the **Zlambach** Formation. The presence of a neritic fauna and the increasing proportion of terrigenous material in the sediments suggests a gradual reduction in depth of water.

II. The Dachstein Reef Complex

The Dachstein Limestones represent a reef complex more than 1200 m thick, stretching from west to east. Throughout late Triassic times, the growth of constructing organisms kept pace exactly with regional subsidence, keeping the reefs close to the surface of the water.

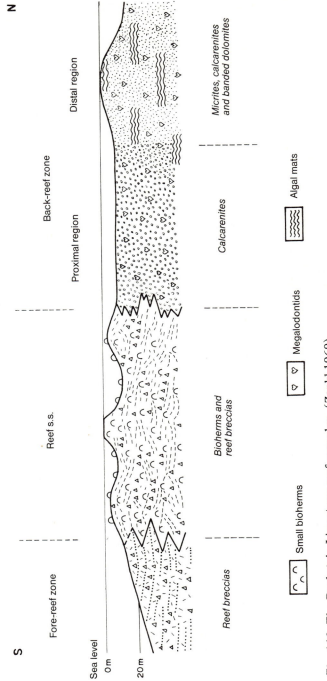

Fig. 113. The Dachstein Limestone reef complex. (Zankl 1969)

From south to north, i.e., from the open sea towards the coast, the Dachstein Limestones can be divided into several facies which lay parallel to the ancient shoreline. There is thus: a fore-reef zone, the reef proper and a back-reef zone (Fig. 113).

1. The Fore-Reef Zone

The Sediment

The fore-reef zone is characterised by **breccias** derived from erosion of the reef. The deposits have a chaotic appearance; they are poorly graded and have no apparent bedding; they are cemented by red micrite. They have clearly been emplaced by flow on the gentle slope from the reef to the deeper parts of the continental shelf.

2. The Reef

a) The Sediment

The reef itself has a width of about 1000 m. It actually consists of a large number of **patch reefs**, each one of which forms an asymmetric dome with a basal diameter less than 5 m and a height of between 0.1 m and 2 m. These organic lenses are built amongst the debris coming from their own destruction, which occurs both mechanically (waves, currents) and through the work of lithophagous organisms. The matrix separating the bioherms consists of poorly rounded fragments, both of organisms (**bioclasts**) and of pre-existing rock (**intraclasts**), which are often graded. Cavities in the sediment have an early infill of **sparite** which helped consolidation. It is often cross-bedded. The matrix forms approximately 90% of the limestones of the reef. All the indicators of a high energy environment are present.

b) The Organisms

The reef fauna, which took part in the construction of the bioherms, is formed of equal parts of **scleractinians** (*Astraeomorpha, Thecosmilia, Montlivaltia,* etc.) and **calcareous sponges** (pharetrones and sphinctozoans) (Fig. 114). They are in their life position. The massive forms occur around the edges of the bioherms, while the branching species prefer the more sheltered central part.

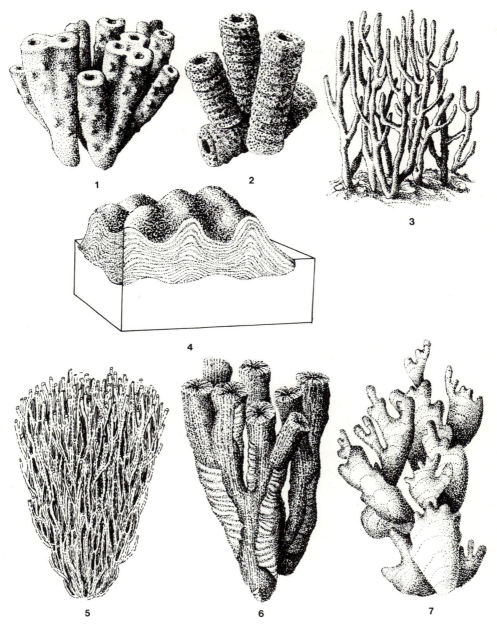

Fig. 114. Fauna and flora of the Hoher Göll reefs. Calcareous sponges: *1 Peronidella*
(× 0.5); *2 Polytholosia* (× 0.5). Coelenterates: *3 Astraeomorpha* (branching form)
(× 0.3); *4 Astraeomorpha* (massive form)(× 0.3); *5 Thecosmilia* (× 0.06); *6 Montlivaltia* (× 0.15); Calcareous algae: *7 Solenopora* (× 0.75). (Zankl 1969)

Scleractinians and calcareous sponges form 75% of the reef builders. The remaining 25% includes calcareous red algae (solenoporaceans), cemented foraminifera, hydrozoans, bryozoans, etc.

Detailed study of the bioherms has shown that growth passes through five stages, following an invariable pattern:
- the bioherm is started by a first generation of scleractinians, sponges and calcareous algae;
- after their death, their skeletons are encrusted by foraminifera and various algae;
- then there comes a third generation of organisms, chiefly calcareous sponges and bryozoans;
- the fourth population is identical to the second;
- the fifth stage of the growth of the bioherm is characterised by calcareous encrustations probably of algal origin *(Spongiostromata).*

These five generations of organisms which succeed each other within the bioherms seem to correspond to a biological cycle. The different species are associated in biocoenoses which are characteristic of each bioherm.

Alongside these fixed, reef-building organisms lives a **sub-reefal** fauna, which contributes actively to the destruction of the bioherms. Numerous species bore into the skeletons (sponges, annelids, bryozoans, bivalves, gastropods, etc.). The green algae also take part. Other forms browse on the calcareous skeletons (crustacea, holothurians, fish). The sub-reefal fauna also adds shells and calcareous skeletons to the sediment, especially those of the benthic foraminifera, molluscs (bivalves and gastropods), brachiopods and echinoderms (sea urchins and crinoids). There is also a sub-reefal fauna of green algae with calcified thalluses (dasyclaceans).

These reef organisms call to mind a **tropical sea** of normal salinity.

3. The Back-Reef Zone

The back-reef zone is several kilometres wide and can be subdivided into a region close to the reef and a region further away from it.

a) The Proximal Region

The proximal region is directly under the influence of the reef. The sediment essentially consists of a well rounded calcarenite, which comes from the breakdown of the calcified thalluses of the dasycladaceans (diplopores) which formed large meadows. The orientation of the grains within the rock itself has suggested that the currents flowed from south to north or from west to east, suggesting that the prevailing wind direction was from the south or southeast.

Besides the green and the red algae (solenoporaceans), other organisms were onkoliths, foraminifera tests, coral fragments and debris of thick-shelled bivalves (megalodontids).

All these features indicate a high energy environment.

b) The Distal Region

In the distal part of the back-reef zone, which is much more widespread than the proximal region, more than 1000 m of fine-grained, regularly bedded limestones were deposited in a quieter environment. This is the **Lofer** facies (Fig. 115).

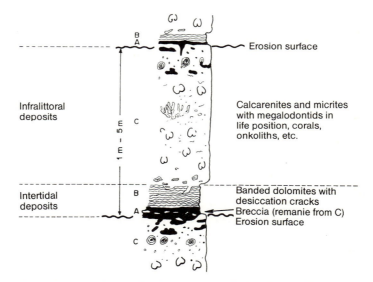

Fig. 115. Sequences in the back-reef zone (Lofer facies) of the Dachstein Limestone. (Modified from Fischer in Zankl 1971)

Each bed, which is several metres thick, represents a three-part **cyclothem**, separated from the underlying and overlying beds by an erosion surface. The lower unit (A) is a calcareous breccia with an argillaceous matrix, reworked from the underlying bed. It is overlain by banded dolomites which were originally stromatolites with a birdseye texture and desiccation cracks; this makes up the second unit of the sequence (B). The third unit (C), thicker than the others, consists of calcarenites and micrites with numerous megalodontids in life position. The openings of the shells often show a preferred orientation, probably under the influence of the currents. The faunal picture is completed by foraminifera, corals, gastropods, echinoderms and various burrowing organisms. The flora includes various calcareous algae.

The lateral extension of the first two units is small. They correspond to the deposits of the **supratidal zone** and of parts of the **intertidal zone** which were subject to long periods of emergence. Algal mats developed in this region and were the origin of the limestones which are laminated on a millimetre scale, rapidly altered to dolomite. The third unit of the sequence, of much greater lateral extent, was deposited in the constantly submerged environment of the **infralittoral zone** where a rich benthos thrived.

The distal region of the back-reef zone thus represents a large **littoral area** scattered with partially emergent mud-banks which was colonised by algal mats. Megalodontids thrived close to a permanent, well-aerated body of water. The succession of cyclothems (approximately 300 for the whole of the Dachstein Limestone) is the result of 1000-year fluctuations of sea level.

III. The Hauptdolomit

Towards the north, the Dachstein reef complexes pass gradually to the carbonates of the Hauptdolomit (Fig. 112), which is, locally, up to 2000 m thick. There are several facies:

— regularly bedded **grey dolomite**, containing benthic animals (bivalves, gastropods) and signs of bioturbation. The sediment is rich in faecal pellets The deposit was laid down in a well-aerated environment, beneath a permanent body of water. These conditions occur in the **infralittoral zone**;

— **banded dolomites** of stromatolitic origin. The rock is a dolomicrite, often with birdseye texture, the pores of which are filled with sparite crystals; they were formed when bubbles of air were imprisoned in the sediment during emergence. Desiccation cracks and streaks of carbon occur. Sometimes plant roots are found in life position in the sediment. The aquatic fauna is very impoverished (foraminifera, ostracodes). The many signs of emergence in this marine area are characteristic of the **shoreface zones**;

— **marly bituminous limestones** rich in remarkably well-preserved fish. The beds are graded and show subaquatic slump structures. This facies represents the infilling of **pools of stagnant water** where reducing conditions favoured the preservation of organic matter.

The Hauptdolomit thus seems to be the product of sedimentation in a large **lagoon** between the reef barrier and land. This lagoon was subject to the rhythm of the tides. Its surface was scattered with algal mats and pools of stagnant water. Elsewhere intertidal conditions occurred.

The dolomitisation of the original sediment occurred very soon after deposition since fragments of dolomite are included in the overlying carbonates. They are therefore syngenetic dolomites.

IV. Conclusion

The reefs of Hoher Göll are markedly different from the classic picture of dome-shaped organic build-ups which separated a lagoon from the open sea. They fit into a broader palaeogeography which has the advantage of showing the lateral extent of different carbonate facies and of presenting a detailed picture of the Mesogean continental platform towards the end of the Trias. Towards the open sea, the sediments consisted of the fine-grained Hallstatt Limestones with a pelagic fauna. The reef area developed in shallower water and was subdivided into a fore-reef zone, the reef sensu stricto and a back-reef zone; it was unusual in being made up of a multitude of small bioherms separated by a mass of biogenic debris. This complex forms the Dachstein Limestone. Close to the shore-line was a vast lagoon subject to tidal influences where the Hauptdolomit was deposited.

Chapter 12 The Holzmaden Bituminous Shale Sea

The village of Holzmaden is situated at the foot of the Schwäbische Alb, about 30 km southeast of Stuttgart (Württemberg) in Germany. The many quarries in this region work the Lower Toarcian bituminous shales or Posidonia Shales. Because of their palaeontological interest, the region around Holzmaden has been declared a protected area.

I. The Sediment

1. Petrology

The Posidonia Shales, which are several metres thick, are actually formed of grey marls separated by a few horizons of limestone.

The marls are rich in **organic matter**, up to as much as 20% by weight of the rock. The extraction of bitumen was formerly undertaken several times during periods of shortage. Pyrite, normally dispersed on a microscopic scale, can also develop into concretions several centimetres in diameter. The pyrite and the organic matter are responsible for the dark colour of the rock.

The activity of the quarries near Holzmaden centres chiefly on working an 18-cm thick shale horizon, the *"Fleins"*. It is harder than the surrounding rock and has a pleasing decorative effect: it is used for facing interior walls in houses.

2. Stratinomy

The marls and the limestones are planar parallel bedded. The individual beds are of a wide lateral extent, but their thickness can vary from one outcrop to another.

The bedding of the marly horizons consists of a succession of fine laminae lying parallel to the stratification; the rock splits easily along them. With a few exceptions, the beds are not bioturbated and no current structure has ever been found.

II. The Fauna

Fossils are found at every level in the Posidonia Shales. Vertebrates and echinoderms are concentrated in the lower part of the succession, while cephalopods are more abundant in the shales at the top. They have no preferred orientation.

1. Burrowing Organisms

Apart from a few well-defined horizons with burrows of the genus *Chondrites,* the Posidonia Shales have no endofauna.

2. Benthic Organisms

Benthic forms are extremely rare. Some ambulant decapod crustaceans *(Proeryon)* have been found near to Holzmaden. Sea urchins *(Cidaris)* and brachiopods are found sporadically. The bivalves *(Posidonia, Inoceramus, Pecten),* which are locally very abundant, clearly come from other biotopes and have been brought into the environment by currents or by floating objects to which they were fixed.

3. Nektonic Organisms

Most of the fauna of the Posidonia Shales consists of organisms which swam in the topmost waters of the sea.

a) Cephalopods

The ammonites belong to four main genera: *Dactylioceras, Lytoceras, Harpoceras* and *Phylloceras.* They can grow to a very large size (several decimeters in diameter). The operculum or *aptychus* is found in place in the body chamber in many shells. Thousands of individuals at different stages of their ontogeny are concentrated at some horizons.

The dibranchiate cephalopods are represented by belemnites and by forms close to squid *(Geoteuthis)* whose soft parts, especially the ink sac, are sometimes fossilised.

b) Fish

Many beautifully preserved fish have been etched out by the technicians of the Hauff Museum in Holzmaden with a goldsmith's care: they provide a vast panorama of the fish faunas of the Jurassic. There are various bony fish, coelacanths and cartilaginous fish. Amongst the bony fish, thick-set forms, whose body was covered by a heavy carapace of large, lozenge-shapes scales *(Dapedius)* co-existed alongside slender forms with tiny scales, the teeth of a carnivore and a more powerful musculature *(Pachycormus)*. The former fed on the bottom in the same way as present-day carp, while the latter actively pursued their prey. *Leptolepis,* the only teleost, was only a few centimetres long and lived in shoals. The sharks are represented by the genus *Hybodus;* the stomach of a specimen preserved in the museum at Holzmaden still contains the remains of around two hundred belemnites which formed its last meal.

Some forms (selaceans, chondrostians and coelacanths) reached a size of 1 to 3 m in length. Clearly they arrived in the depositional environment as **adults**; in general, juveniles are rare. This suggests that their development took place elsewhere, in more favourable waters, and that only the adults penetrated the Holzmaden area.

c) Reptiles

The reptiles are the most spectacular fossils in the Holzmaden horizon. To date, more than 300 complete skeletons of ichthyosaurs have been extracted, and have enriched museums the world over.

The **ichthyosaurs**, which are several metres long, are reptiles remarkably adapted to swimming (Fig. 116A). We know the outline of their neckless, fusiform body and of their fins from some specimens which have the soft parts preserved. The vertical heterocercal tail fin propelled the animal forwards, while its paired paddle-shaped fins supported it. Ichthyosaurs swam close to the surface of the water to chase their prey, which consisted mainly of dibranchiate cephalopods, whose hooks have been found in the stomachs of many reptiles. Fossil **embryos** inside the body of their mother indicate that ichthyosaurs were oviparous. This reproductive adaptation dispensed with the need for these marine animals to return to dry land to lay their eggs.

Plesiosaurs are much rarer in the Holzmaden horizon (Fig. 116B). They must have lived elsewhere, probably in the open sea. These huge reptiles propelled themselves forwards by means of paired paddles which were attached to powerful scapular and pelvic girdles. The poorly developed tail acted as a rudder. A tiny head whose jaws were studded with

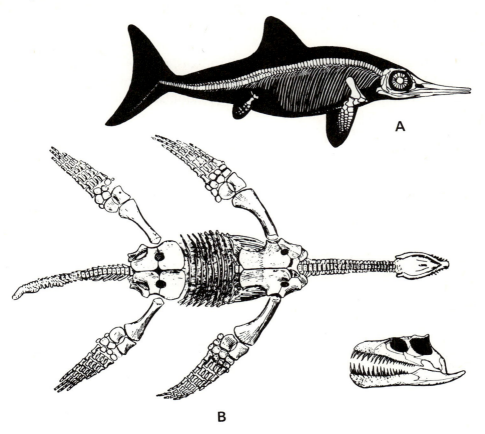

Fig. 116 A, B. Toarcian reptiles from Holzmaden. A ichthyosaur *(Stenopterygius)*. (Grasse 1970); B plesiosaur (skeleton of *Rhomaleosaurus* and head of *Plesiosaurus*). (Kuhn 1968)

sharp teeth was held at the end of a long and flexible neck. Unlike the ichthyosaurs, the plesiosaurs had to lie in wait for their prey to pass close by and then seize it with the help of the mobile anterior part of their body, like present-day swans.

A third group of reptiles is represented by the **steneosaurs,** which are related to the crocodiles. The shape of their body, protected by an armour of bony plates and drawn out into a tapering muzzle, is reminiscent of the gavials (Ganges "crocodiles"). Some specimens contain **gastroliths,** stones collected on the shore by the animal to reinforce the grinding action of the stomach on its food. The Holzmaden steneosaurs are very large. Juveniles are rare since the first stages of growth take place close to land.

4. Pseudoplanktonic Organisms

Crinoids *(Seirocrinus)* and bivalves (*Inoceramus, Posidonia, Monotis,* etc.) normally live fixed to the sea bed. However, in the bituminous shales, their abundance is associated with a pseudoplanktonic mode of life. These animals grow on objects floating near the surface of the water: tree trunks or empty shells. When these became heavy and gradually sank to the poorly oxygenated deeper waters, they dragged the animals to their death. On a plaque preserved at the Institute of Geology in Tübingen, Seilacher, Drozdzewski and Haude have reconstructed the different stages of burial of a colony of crinoids fixed to a floating support (Fig. 117).

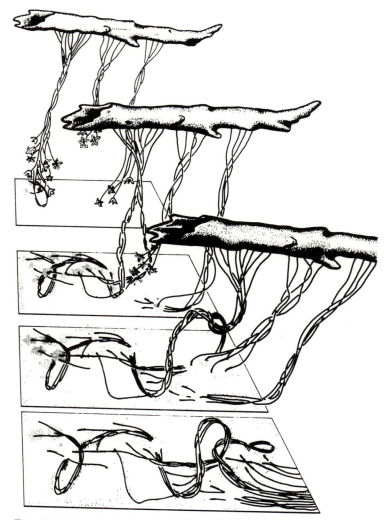

Fig. 117. Pseudoplanktonic mode of life and different stages of burial of a crinoid colony *(Seriocrinus)* from the Toarcian of Holzmaden. (Seilacher et al. 1968)

An even more striking palaeontological specimen is on exhibition at the Hauff Museum in Holzmaden: a slab of shale 18 X 6 m showing a 12-m-long tree trunk entirely covered by a multitude of shells of *Inoceramus* with some crinoids mingling their supple stems and gracefully displaying their calyces.

5. Flying Organisms

Some skeletons of flying reptiles (**pterosaurs)** have been excavated at Holzmaden. They suggest the proximity of land.

III. The Flora

The plant remains in the Posidonia Shales sometimes reach a considerable size. They consist mostly of floating tree trunks which have been transported a long way before burial. Apart from a few cycad and conifer *(Plagiophyllum)* branches, the plants are unidentifiable.

The microflora consists of pollen and diatoms.

IV. Fossilisation

1. Anaerobic Conditions

The extreme rareness of burrowing and benthic organisms, compared to the variety and abundance of planktonic and pseudoplanktonic forms, indicates that the sea floor muds and the overlying waters were unsuitable for living organisms. The abundance of pyrite and of undegraded organic matter in the sediment proves that the environment was reducing. Only the uppermost layers of the water were sufficiently oxygenated to allow the spread of life.

These anaerobic conditions, hostile to the growth of saprophagous organisms, were very favourable for the fossilisation of soft parts. Moreover, they slowed down the bacterial decomposition which normally produced a large amount of gas in the body and caused it to float, disarticulating the skeleton. In this special situation, organisms stranded on the bottom did not rise to the surface of the water again. Their skeletons were thus preserved whole. This type of fossilisation, though spectacular, nevertheless remains exceptional.

The characteristics of the environment were not uniform throughout the whole of the deposition of the Posidonia Shales. Some horizons contain *Chondrites* burrows, others have large accumulations of shells. This can be related to a temporary oxygenation of the bottom waters and to the action of local currents.

Recent work has radically altered such an interpretation.

By measuring the orientation of different fossils in the Posidonia Shales (bivalves, ammonites, belemnites, ichthyosaurs and plant remains), K. Brenner (1976) has shown that N-S or S-N bottom currents were active throughout the whole time of deposition. These currents, which were relatively fast (20 cm/s), followed a large N-S marine depression which paralleled the shelf.

Kauffman (1978) has shown that the Lower Toarcian sedimentary succession includes a marine transgression and regression. The maximum transgression occurred during the deposition of the coccolith-rich limestones at the top of Middle Lias ϵ (Schlaken). It represents the time when the availability of free oxygen was at its lowest. On either side of this horizon are facies which were more favourable to benthic faunas. These consisted of mud eaters, whose burrows are abundant at the top and bottom of the formation *(Chondrites, Spongeliomorpha)* as well as serpulids, brachiopods and bivalves *(Posidonia, Ostrea,* etc.). The latter lived fixed to objects which had been stranded on the bottom, such as the shells of large ammonites, which acted as little islands keeping their inhabitants raised above the reducing zone. Kauffman suggests that the boundary between the oxidising and reducing zones continually oscillated within the sediment itself and the immediately overlying water (Fig. 118). An algal mat probably sheltered this zone from the action of the bottom currents.

2. Burial

Depending on their density, dead organisms sank more or less deeply into the soft mud. They were then gradually covered by clay particles which gave rise to the bituminous shales. But the **rate of sedimentation** was low, so that the part of the skeleton which remained above the mud became altered. This is why the extraction and preparation of fossils is always done from their lower side, which alone remains intact.

In contrast, the limestones, which are poorer in organic matter, must have been deposited more rapidly than the marls.

Fig. 118. Small-scale fluctuations of aerobic $(+O_2)$-anaerobic to oxygen-depleted $(-O_2)$ boundary on the sea floor of the bitumunous shales of Holzmaden, and response of benthic biota. (Kauffman 1978)

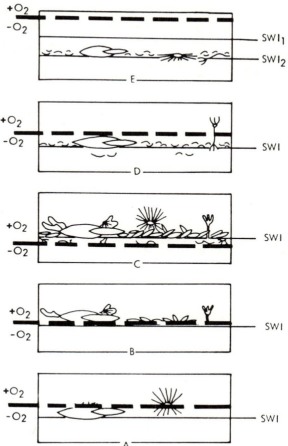

3. Diagenesis

The stenohaline nature of the organisms (cephalopods, echinoderms) is characteristic of an open marine environment. The regularity of the bedding confirms that the body of water was large.

Floating tree trunks came from far-distant emergent lands where the early growth stages of the steneosaurs and the flying reptiles also took place. Life was only possible in the aerated upper parts of the water. The fauna, however, remained of low diversity because living conditions were rigorous. Amongst the vertebrates, only the ichthyosaurs seem to have lived and reproduced in the same place as they were buried. The cephalopods and the crinoids had undergone considerable transport by marine currents before they sank to the bottom. Thus the assemblage is essentially a **thanatocoenosis**. One should not be deceived by the beautiful specimens in the collections: most of the specimens collected are incomplete or broken; whole examples are an exception.

In early Toarcian times, the Holzmaden region was a marine bay where **stagnant waters** allowed fine particles to settle out and created conditions hostile to the development of life close to the sea floor; these conditions were, however, favourable for the fossilisation of organic matter. Similar conditions occur at the bottom of the Black Sea at the present day. However, the euxinic nature of the sedimentary environment was often disturbed by currents which orientated the dead organisms lying on the bottom, aerated the water and allowed temporary colonisation by a benthic fauna.

The action of the surface currents, continued over millions of years, collected the bodies of nektonic or pseudoplanktonic animals which lived in the surrounding seas together. The Holzmaden horizon is thus a gigantic **marine graveyard.**

Chapter 13   The Solnhofen Lagoon

For several centuries, the lithographic limestones *("Plattenkalke")* out-cropping between Munich and Nuremberg in northwest Bavaria have been worked for flagstones. In this formation, which is early Portlandian in age, are some very famous fossiliferous horizons, of which the best known are those of Solnhofen and Eichstätt.

I. The Sediment

1. Petrology

The limestones around Solnhofen are 96%–98% **micrite**, in other words, microcrystalline calcite. This is partly because of their coccolithic origin. The particularly fine and uniform grain of the rock has made it a prime choice for lithography.

At outcrop, the limestone is a beautiful beige colour, ornamented by dendritic manganese oxide along joints and cracks. When it is fresh, the sediment is grey-blue because of the presence of iron sulphide and it is the oxidation of these sulphides on contact with the air that turns it to its characteristic yellow.

2. Stratinomy

a) Bedding

In the Solnhofen region, the lower Portlandian limestones form the infill of a system of **little basins**, separated from each other by sponge reefs. A **reef barrier** which perceptibly follows the present-day course of the Danube, isolated this group of deposits from the open sea to the south. The thickness of the series changes from a few metres to more than 60 m as one goes from the edge to the centre of the basins. Moreover, the beds increase in thickness and limestones become dominant over the marly horizons.

These deposits outcrop as an undisturbed, regular succession of beds of limestone *("Flinzen")* and more marly layers *("Fäulen")*. The thickness of each bed varies from a few millimetres to 30 cm. They have a very great lateral extent: some beds can be followed for more than 8 km. The limestone and marl beds consist of a pile of fine, millimetre-scale, laminae with bedding planes which are sub-horizontal and generally smooth.

b) Sedimentary Structures

The regular succession of the bedding is locally disturbed by subaquatic slides and load casts. The existence of currents can only be proved in the eastern part of the outcrop, where one can see graded beds, flute casts and groove casts. Elsewhere, the orientation of some fossils (fish) and traces of ammonite shells rolling indicate only moderate currents (Fig. 119). The rare ripple marks are of the oscillation type.

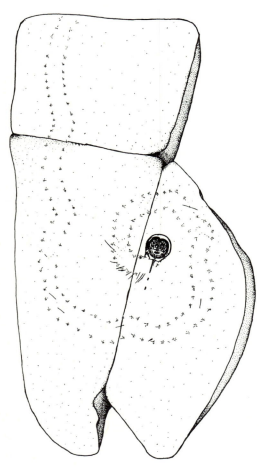

Fig. 119. Dying traces of a limulid on a bedding plane in the Solnhofen lithographic limestones (from a slab preserved in the Museum at Solnhofen)

The hypothesis of a temporary emergence of the environment of deposition has not been confirmed, since the desiccation cracks described by some authors have been ascribed to submarine shrinkage.

II. The Fauna

More than 600 species of animal have been listed from the lithographic limestones. Most occur at specific levels in the formation. However, the variety and the admirable preservation of the organisms should not conceal the low density of fossils. Only patient research undertaken for more than a hundred years by a large number of collectors has accumulated such a wealth of palaeontological information.

1. Burrowing Organisms

The fine laminations of the sediment are not bioturbated; from this, one deduces the absence of burrowing organisms.

2. Benthic Organisms

Compared to the whole fauna, animals which lived in contact with the bottom are rare.

The **macrofauna** includes polychaete worms *(Eunicites)* whose jaw apparatus and parapodial acicula are sometimes preserved, crustaceans *(Aeger, Eryma, Eryon,* etc.) with remarkably intact appendages, limulids *(Mesolimulus)* (Fig. 118) and echinoderms (ophiuroids, sea urchins, holothurians). As a general rule, the limulids and crustacea are represented by juveniles; adults are rare.

The **microfauna** consists of foraminifera (notosarids, miliolids) and ostracodes.

3. Nektonic Organisms

a) Coelenterates

The exceptional quality of the fossilisation in the lithographic limestones is illustrated by the preservation of the imprints of **medusids**. Although they are rare forms, often unidentifiable, some specimens show features of the musculature of the bell and details of its morphology *(Rhizostomites).*

b) Cephalopods

About 30 species of ammonite have been recorded in the Solnhofen Limestones (*Perisphinctes, Oppelia, Aspidoceras,* etc.). Many specimens still retain their operculum (**aptychus**) inside the body chamber. Some show the remains of soft parts.

The dibranchiate cephalopods include **belemnites** and forms close to squid which can be up to 1.50 m long. In some specimens, the ink sacs and the arms with their hooks have been fossilised *(Acanthoteuthis).*

c) Fish

An impressive list of 150 species of fish has been recorded from the lithographic limestones, including cartilaginous fish, coelacanths and bony fish.

The cartilaginous fish are represented by the sharks and the chimerids. However, the richness of the Solnhofen fish fauna is particularly striking in the bony fish where species of very different shape and size occur together (*Lepidotus* can be more than 2 m long). *Leptolepis,* a teleost a few centimetres long, is by far the commonest fish. Abundant material containing examples of both juveniles and adults of *Aspidorhynchus* has made it possible to work out the complete series of morphological changes which took place during ontogeny. Because of the excellent preservation, it has been possible to study the stomach contents of several other species.

d) Reptiles

Marine reptiles are represented at Solnhofen by rare traces of **ichthyosaurs**, by **turtles** and by crocodiles. Some turtles *(Eurysternum)* have a flat, incompletely ossified, carapace. This lightening of the body is an adaptation for life in the open sea. Pelagic forms also exist amongst the crocodiles which develop a dorsal and a caudal fin *(Geosaurus).*

4. Planktonic Organisms

A major part of the limestone consists of **coccolithophorid** tests; these flagellates are part of the nannoplankton.

Saccocoma, a small, stemless crinoid, is the only really common fossil in the Solnhofen horizon. Millions of individuals have been washed into the environment of deposition.

Higher crustacean (palinurids) **phyllosomid larvae**, which are planktonic forms disseminated by the waves, also occur.

5. Pseudoplanktonic Organisms

The rare bivalves found in the Solnhofen Limestone arrived in the environment of deposition by attaching themselves to floating objects: algae or shells. For example, juvenile oysters *(Ostrea)* sometimes encrust ammonite shells.

6. Terrestrial Organisms

Animals which lived on dry land are extremely rare. A rhynchocephalid reptile, *Homeosaurus,* is known from about 20 skeletons. Like the present day lizards, this vertebrate seems to have been able to regenerate its tail, which have been found preserved in the stomachs of fish several times.

7. Flying Organisms

The abundance of flying organisms is connected with the proximity of dry land.

a) Insects

A crowd of insects (83 genera, 141 species) has been discovered in the Solnhofen Limestones: ephemerids, libellulids, blattids, orthopterids, coleopterids, nevropterids, lepidopterids, dipterids, hymenopterids and hemipterids. However, the number of individuals is generally low. Some orthopterids *(Chresmoda, Elcana)* have feet especially adapted for walking on water; the musculature of these feet is sometimes preserved.

b) Pterosaurs

The flying reptiles of Solnhofen *(Pterodactylus, Rhamphorhynchus)* have greatly contributed to our knowledge of this group. Their state of preservation is remarkable: it has been possible to study their flight membranes, their body hair, the webbing on their hind feet and the contents of their stomachs. Their size varies between that of a sparrow and that of an eagle. These animals, which were excellent at flying, were also good divers. They ate insects and fish.

c) Birds

The most famous fossil from the Solnhofen horizon is *Archaeopteryx,* a bird which still retained many reptilian features, in particular teeth and a tail. Five specimens are known at the present time; the feathers are preserved on three of them.

III. The Flora

Plants are very rare at Solnhofen: there are some ferns, some ginkgos *(Baiera),* some conifers *(Brachyphyllum)* and various plant remains attributed to the algae. They indicate a **tropical climate**.

IV. Evidence of Biological Activity

1. Dying Tracks

The most striking examples of evidence of activity found at Solnhofen are the locomotion traces found directly related to the body of the animals that made them (Fig. 119). These traces show the line of the last movement of the animal, immediately before its death − real "agony tracks". Some **limulids** have been preserved like this, at the end of a long trail of footprints and telson scratches. Around the body itself, the many marks left by the telson indicate that the animal thrashed about before dying. **Crustaceans** *(Mecochinus)* have been found in similar situations.

2. Locomotion Traces

Invertebrates have left a variety of locomotion marks: ammonites, dibranchiate cephalopods, crustacea, echinoderms, etc. Some marks have been produced by dead animals whose bodies or shells have scratched the sediment (medusids, ammonites).

3. Coprolites and Stomach Pellets

Coprolites of various shapes and sizes have been produced by fish or cephalopods *(Lumbricaria)*. Stomach pellets containing fragments of fish, cephalopods and crinoids have been attributed to fish and crocodiles.

V. Fossilisation

The state of preservation of the fossils at Solnhofen is admirable: moulds of medusids and feathers, the complete bodies of annelids, crustacea and insects with their appendages still intact, etc., all produced by an exeptional coincidence of conditions.

1. Water Movement

Apart from in the eastern part of the basin, only moderate, episodic, currents occurred during the deposition of the lithographic limestones. Their existence can be deduced from the occasional orientation of fish or ophiurids and tracks of rolling ammonite shells (Fig. 120). A quiet environment is implied by the articulated skeletons and shells and the preservation of aptychi lying concave-up. In general, the fine grain of the sediment and the beautiful regularity of the bedding also reflect a **low energy hydrodynamic regime**.

Fig. 120. Orientation of 266 fish *(Leptolepis)* on the surface of a slab of Solnhofen lithographic limestone *(points of the arrows* indicate the position of the animals' heads). (Janicke 1969)

2. Rate of Sedimentation

Some fossils are orientated obliquely to the bedding. Belemnite rostra and plants can cut across several laminae of the same bed (Fig. 121). Ammonite shells have been found standing vertical. These observations suggest rapid deposition. The presence of load casts also suggests a high

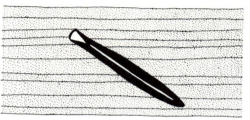

Fig. 121. Belemnite rostrum lying obliquely in a bed of lithographic lime- stone (indicator of a high rate of sedi- mentation)

rate of sedimentation, with the mud still being full of water when the overlying bed was deposited.

Some beds of limestone could have been deposited within a few days. These conditions are eminently suitable for fossilisation.

3. Anaerobic Conditions

The decomposition of organic matter was slowed down by the reducing nature of the mud. This is suggested by the presence of iron sulphide in unaltered rocks. The fine-grained nature of the sediment, the regularity of the bedding, the absence of burrowing organisms and the rarity of benthos confirm that the water was stagnant.

The dying traces are thus easily explained: when the animals entered a restricted environment, with a reduced oxygen level, they quickly died when they made contact with the bottom.

4. Salinity of the Water

According to H. Keupp (1977), the remarkable preservation of the Soln- hofen fauna can be explained by periodic hypersalinity of the water in the environment of deposition. Such conditions led to mass mortality of the benthic fauna and at the same time helped their fossilisation. Sub- sequently, a mat of blue-green algae protected their remains, while rapid precipitation of a fine calcareous mud allowed the preservation of minute anatomical detail of the organisms.

5. Fineness of the Sediment

The fine-grained nature of the lithographic limestones is ideal for retain- ing very delicate morphological detail.

6. Diagenesis

During **compaction**, the sediment lost between four and eight times of its original volume and the fossils were flattened.

Some macrofossils (ammonites, crustacea, fish) are perched on a **"pedestal"** formed on the upper surface of the beds. These animals had been rapidly buried with their soft parts or cavities full with water. When the density of the surrounding sediment increased during compaction, the fossil, which remained lighter, tried to rise upwards. Compensation took place within the sediment, causing the laminae which form the "pedestal" to be deformed (Fig. 122).

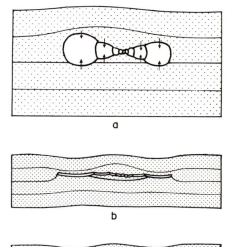

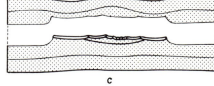

Fig. 122a–c. Formation of a "pedestal" beneath an ammonite in the lithographic limestones by flattening of the shell and compaction of the sediment. (*Arrows* indicate the directions of the pressures on the fossil after burial). (Van Straaten 1971)

VI. The Environment

The Solnhofen lithographic limestones lie in a huge basin some 100 km long and 30 km wide, which was subdivided into smaller basins by sponge reefs. A massive reef barrier protected it from the open sea (Fig. 123).

In this **lagoon**, a body of quiet water, the depth of which has been estimated to have been about 50 m on average, was subjected to intense evaporation in a tropical climate. The waters were poorly aerated and became stagnant; the oxygen content decreased. Reducing conditions were developed at depth with the production of hydrogen sulphide: this gave rise to the iron sulphide in the sediment. An environment such as this is lethal to an aquatic fauna.

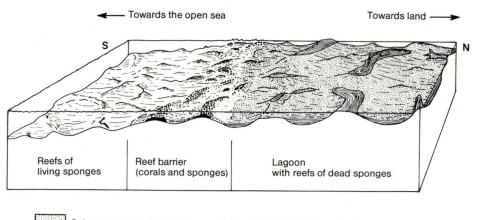

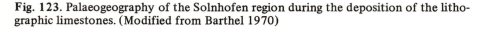 Calcareous mud (future lithographic limestone) Turbidity currents

Fig. 123. Palaeogeography of the Solnhofen region during the deposition of the lithographic limestones. (Modified from Barthel 1970)

Periodically, the lagoon was the site of carbonate sedimentation which rapidly buried the organisms. Calcareous mud started to move downslope from the sea or the nearby reefs and was carried in dense, slow suspension currents (**turbidity currents**) towards the centre of the basins where it was deposited in regular sheets of wide lateral extent. This produced the bedded limestone ("Flinzen"). The marly layers ("Faulen") were deposited between two successive flows. This process, repeated many times, produced the beautiful regularity of the bedding. As they moved, the currents caught up the animals which lived close to the shores, near the reefs or on dry land. They quickly died in the inhospitable bottom waters because of asphyxiation or suffocation in the mud, as is shown by their dying traces. The slow movement of the currents in the denser waters of the deep parts of the lagoon explains the rarity of erosion marks; the arthropods which were caught up in the currents tended to be **juveniles** because they offered less resistance than the larger adults.

According to Keupp (1977), the body of water was subjected to an alternation between hypersaline and normal marine conditions many times. When the water was supersaturated, blue-green algae proliferated and induced abundant precipitation of calcium carbonate. When the water was of normal salinity, the algal mat was resorbed and planktonic and nektonic organisms accumulated on the cleared surface, where benthic forms were also able to live for a while. A new episode of stagnation and supersaturation resulted in the death of the benthos and the return of the blue-green algae. By covering the skeletons and dead animals, the algal carpet protected them from alteration and ensured their fossilisation in a micritic mud.

In one bed of limestone, a single primary sequence corresponding to one period of restriction was on average 0.5 mm thick. This period of restriction can only have been a few years long.

The rhythmic nature of the sedimentation is considered to be the result of winds, which occasionally disturbed the layering in the waters of the lagoon.

According to some authors, the mass mortality of the aquatic fauna must have been periodically induced by **"red tides"** following a sharp bloom of phytoplankton which produced toxins and lowered the dissolved oxygen content of the water. In this case, this role was played by the coccolithophorids whose shells are abundant in the limestones.

The organisms which ended up either moribund or dead in the lagoon found ideal conditions for fossilisation when they were buried by fine-grained reducing sediment. This **thanatocoenosis** has produced one of the best fossiliferous horizons in the world. Thus, for 250,000 years, the Solnhofen lagoon acted as a fatal trap for the fauna of the neighbouring shores.

Chapter 14   The Shores of the Auversian Sea

At the beginning of the late Eocene, in **Auversian** or **early Bartonian** times, a marine transgression flooded the Paris Basin from the west. It formed a gulf which reached as far as Champagne to the east and which was separated from the Belgian basin by the Artois ridge. This sea was surrounded by a complex of lakes and lagoons (Fig. 124).

The marine facies of the Auversian outcrops in many sandpits to the north of Paris, especially in the departments of Oise (Ermenonville) and Val d'Oise (Attainville, Auvers sur Oise, Le Guépelle, Ronquerolles, etc.).

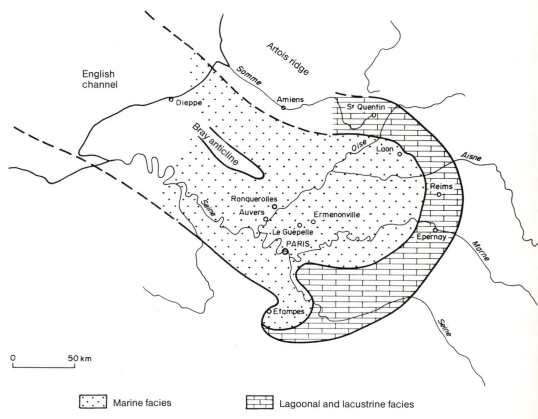

Fig. 124. Extent of the Auversian gulf. (Pomerol 1973)

I. The Sediment

1. Petrology

The marine facies of the Auversian chiefly consists of yellow, grey or white calcareous **sands**, sometimes cemented to sandstones. Locally there are intercalations of clay or limestone. The sands can be glauconitic and the abundant shell fragments can make the rock a biocalcarenite. The average thickness is about 25 m.

a) Grain Size

In general, the Auversian sands are fine and well-graded. Statistically, the mean grain diameter (Q_2) increases from the west (0.15 mm in Parisis) towards the east (0.25–0.3 mm in Brie). Moreover, the grading becomes poorer and the frequency curves are more often bimodal as one goes in that direction. This lateral change in grain size suggests that the detrital material probably came from the east.

In Valois and Tardenois, the very white sands in the upper part of the formation have a high proportion of matt, oval grains indicating an **aeolian** influence on their shape.

Locally, pebbles of the Palaeozoics of the Ardennes, the Cretaceous and the Lutetian can be found. Many fossils have been reworked from the lower Eocene and even from the Cretaceous.

b) Heavy Minerals

The heavy mineral suite of the Auversian sands contains a background of ubiquitous minerals common to all the Eocene of the Paris Basin (tourmaline, zircon, rutile, garnet, andalusite, staurolite, etc.) on which a peak of tourmaline is superimposed. The content of this latter mineral increases regularly from the west towards the east. Pomerol has shown that the bulk of the common minerals come from the reworking of the Cretaceous and Eocene sands of the Paris Basin, while the tourmaline comes from the Palaeozoic and Triassic of the Ardennes and the Vosges.

c) Organic Matter

Some horizons of fine sand contain between 2% and 10% organic carbon. Doubtless this comes from plant debris whose nature is still unknown (algae?).

2. Stratinomy

A distinctive characteristic of the Auversian deposits is their great lateral facies variation. Sections change from one outcrop to another, and it is not rare to see a sand passing into a shell bed or a clay even within one quarry.

a) Bedding

The bedding of the Auversian sands is often lenticular; bed thickness varies from 0.20 m to 2 m.

In the fine sands, bedding is frequently horizontal, while coarser sediments are cross-bedded. The presence of pebbles and broken shells confirms that the coarser horizons were transported by stronger currents.

The groups of steeply dipping sedimentary units are interpreted as **meander bar** deposits formed on the banks of tidal channels. The sands are fine-grained, the shells unworn and the bivalves are frequently preserved in position of life.

The groups of crossbeds with bases which are sometimes concave-up are attributed to **aeolian dunes**. The sediment is fine and very well-graded; it contains neither shells nor pebbles and has no marly intercalations.

The unconsolidated nature of the sediment makes palaeocurrent measurements and the observation of sedimentary structures difficult.

b) Beach Rock

In some outcrops of the Auversian, slabs of calcareous sandstone several metres thick can be found. The well-defined upper surface is horizontal, unlike the lower surface which is irregular and less well cemented. These indurated horizons cut the cross bedding of the sands. They are interpreted as **beach rock** which develops within present-day littoral sands as a result of the precipitation of $CaCO_3$ in the contact zone between sea water and fresh ground water. They are particularly common in a tropical climate. These local indurations of the sediment are resistant to erosion and are the origin of sandy nodules.

c) Palaeosols

Dark coloured sands and sandstones with a high organic content and root systems (Fig. 125) occur at the top of the Auversian. Within them one can sometimes recognise two horizons:

Fig. 125. Plant roots in
an Auversian palaeosol
at Ronquerolles.(Plaziat
1971)

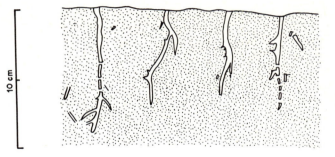

— a pale grey **leached horizon,** 10 to 20 cm thick;
— a brownish black **accumulation horizon,** rich in humus and iron, which
   passes gradually to a lighter-coloured horizon with plant roots.

These palaeosols have the features of podsols.

   Going from west to east, beach sands are replaced by palaeosols, indicating that one is approaching land.

II. The Fossils

For a long time, the Auversian sands have been famous for the abundance
and diversity of their fossils which make up more than 50% of the rock
by weight in the shell beds. The species lists vary greatly from one outcrop to another.

1. The Flora

The flora is mainly represented by the remains of encrusting dasycladacian
green algae *(Dactylopora).* Characian oogonia have been reported in the
lagoonal facies.

2. The Fauna

a) The Microfauna

At the present time, 125 species of **foraminifera** have been described of
which 56 come from the type locality at Le Guépelle. They are benthic
forms belonging to the miliolids, buliminids, polymorphinids and num-
mulids *(Nummulites variolarius)* etc. Their diversity decreases from the
western to the eastern part of the Auversian gulf.

   Ostracodes are abundant in and characteristic of the littoral facies.

b) The Macrofauna

i) Benthic Organisms

Benthic organisms form the main part of the Auversian fauna, even being as abundant as 10,000 individuals per kilogram sample. They belong to different groups: the annelids (serpulids), bryozoans, scleractinians, scaphopods, gastropods, bivalves, echinoderms and crustacea.

The dominant group is the molluscs with more than 200 species listed. The relative abundance of bivalves and gastropods varies from one horizon to another. At Le Guépelle, for example, gastropods are seven times more common than bivalves. At Ronquerolles, on the other hand, the latter are dominant.

The bivalves include burrowing forms (*Teredo, Solen, Cardium, Corbula,* etc.) or fixed forms (*Chama, Ostrea,* etc.). The shells are often thick (*Meretrix, Cardita,* etc.) and still show traces of their original colouration.

The gastropods include carnivorous forms *(Cerithium, Natica).* These are doubtless partially responsible for the holes seen on many shells. Most species, however, seem to have been herbivorous (*Bayania, Rissoa, Ringicula,* etc.). In some horizons, gastropods especially tolerant of salinity variations, or even completely freshwater conditions, are abundant (*Hydrobia, Planorbis,* etc.).

The crustacea are represented by the claws of burrowing decapods *(Calianassa).*

ii) Nektonic Organisms

Fish are represented by otoliths and teeth *(Lamna).*

iii) Terrestrial Organisms

The outcrop at Le Guépelle has yielded **tortoise** scutes and a **mammal** fauna of perissodactyl herbivores (*Lophiodon, Palaeotherium,* etc.) and some carnivores *(Hyaenodon).*

The intercalation of a continental fauna between the marine horizons is most important in Eocene stratigraphic correlation.

III. Environmental Factors

1. Salinity

The diversity of the organisms and the presence of stenohaline groups (bryozoans, scleractinians, echinoderms, etc.) is characteristic of **marine faunas**. Many genera of the foraminfera, gastropods and bivalves still live in modern seas.

Some horizons have an impoverished fauna, where the shells are noticeably smaller than average. Moreover, euryhaline species become more abundant. This indicates a lowering of salinity such as occurs in **lagoons**.

Other horizons yield **freshwater** gastropods, sometimes mixed with the bones of terrestrial mammals. These deposits are clearly the result of the flooding of a small coastal river.

To summarise, the Auversian fauna is typical of a marine environment; this is confirmed by the presence of **glauconite**. However, frequent episodes of low salinity caused by the introduction of freshwater had their effect on the nature of the animal and plant populations.

2. Bathymetry

There are many signs that the water was shallow, including the beach rock and the frequent cross-bedding. Moreover, the occurrence of palaeosols in the eastern outcrops indicates the proximity of dry land.

The composition of the fauna, of which many genera still live in littoral waters, reflects a shallow water biotope. In the beaches, decapod crustaceans made dwelling burrows and boring molluscs made holes. Most of the gastropods lived in the intertidal meadows.

The Auversian sands with their rich shelly fauna correspond to **marine beach deposits** where the water was barely more than a few tens of metres deep.

3. Temperature

The O_{16}/O_{18} isotope ratio in the carbonate of various marine mollusc shells (*Meretrix, Corbula,* etc.) suggests that the mean temperature of the water in which the animals lived was around 25°C. The warmth of the sea is confirmed by the presence of scleractinians and by the thickness of the shells (foraminifera, molluscs). In Auversian times, the Paris Basin must have had a tropical climate.

4. The Hydrodynamic Regime and Associations of Organisms

a) Thanatocoenoses

Many cross bedded coarse sands yield shells which are mostly broken or damaged. They are sometimes concentrated in pockets or sorted according to size. Bivalve shells are typically concave-down. In these horizons, the fauna is at its maximum diversity and density (more than 660 individuals have been counted in 1 g of *Nummulites variolarius* sands!). They accumulated in places favoured by marine currents which were carrying empty shells from different biotopes. These concentrations of dead shells are thanatocoenoses and eventually form **faluns** (shell beds). Sometimes only some groups are allochthonous. Thus, in the Ronquerolles horizon, the gastropods show signs of severe wear while the shells of bivalves are unaffected and the animals must have lived near the place where they are fossilised.

b) Palaeobiocoenoses

The fine sands generally have a less diverse fauna. Even though they are fragile, the shells are intact and signs of wear are minimal. Juveniles and adults typically coexist in these horizons. *Ostrea* shells are found complete with their attachment areas. Moreover, the presence of burrows and of bivalves in life position proves that the fauna is autochthonous. Despite their cross bedding, these horizons were deposited in water which was not very rough, so that most animals were fossilised in the environment in which they had lived. These associations may be called palaeobiocenoses.

c) The Submarine Meadows

Some very fossiliferous, fine-grained, well graded sands are distinctive because of their high organic carbon content (2%–10% of the rock by weight), which probably comes from aquatic plants. The dense fauna has a low diversity. Miliolids and herbivorous gastropods are numerous and the shells of bivalves are still articulated. This is reminiscent of the biocoenoses of present-day Mediterranean submarine meadows.

d) Lagoons

Fine sands and green clays with attapulgite and sepiolite represent sedimentation in quiet, somewhat restricted water. The fauna consists of euryhaline or freshwater species. The ostracods and some foraminifera have shells which are thin and unornamented and the molluscs are often dwarfed. These features are characteristic of lagoonal environments.

IV. Conclusion

The palaeontological and sedimentological characteristics of the Auversian sands are those of **neritic environments receiving terrigenous sedimentation.** As one goes from west to east, one approaches the shore. The closeness to land is indicated by a series of changes: the sediments become coarser, the tourmaline content rises, euryhaline organisms become more important, the microfauna decreases, beach rock then soils start to appear, and so on.

The geographical distribution of outcrops suggests that the Auversian sea formed a gulf elongated E-W within the Paris Basin. The different facies recognised in the stage are related to the many environments of deposition in a low-lying tropical area with changing bathymetry.**Littoral bars** formed above wave base. In this high energy environment, coarse sands and fossils accumulated to form thanatocoenoses. Behind these natural barriers, quieter environments with marine meadows and beaches with burrowing crustaceans and bivalves was established. As it became more restricted and fresh water flowed in, this environment changed to one of lagoons and lakes. Soils and aeolian dunes developed on the emergent landmasses, where mammals and freshwater gastropods thrived.

As a result of the movement of the littoral bars, these various facies were closely interbedded, resulting in an infinite variety of deposits and conditions of life. The growing importance of continental influences towards the top of the stage reflects the marine regression which ended the Auversian episode in the Paris Basin.

Chapter 15   The Acheulian Hunters' Cave at Le Lazaret

The cave of Le Lazaret is hidden on the slopes of Mont Boron, close to the town of Nice. Its opening lies less than 100 m from the present-day shore of the Mediterranean; it is a natural cavity about 40 m long by 20 m wide formed by karstic erosion in an early Quaternary conglomerate, which itself is a fissure infilling in Jurassic limestones. The Quaternary sedimentary infill of the cave is more than 7 m thick; the upper part of this infill was deposited between the end of the third and last phase of the **Riss glaciation** and the beginning of the **Würm glaciation**. It was excavated in 1967 by a team headed by H. de Lumley.

I. Excavation Techniques

The principle of the excavation of prehistoric sites is to remove successive lithological horizons over a wide area, collecting information about the sediment, the fauna, the flora and man from each level.

To begin with, a **vertical scale** to relate the various layers to one another and a **numbered grid** to locate the position and orientation of each object discovered are set up (Fig. 128). The squares of the grid, which have sides of 1 m, are called **zones**. The sedimentary layers in each zone are removed one after another by means of a spatula or a paintbrush. The sediment which has been collected is used for sedimentological and palynological analysis and, after sieving, for micropalaeontological studies. The fragile bones are impregnated and strengthened as they are freed.

All this sedimentological, palaeontological and palaeoethnographical information is then recorded, level by level, on a **plan** at a scale of 1:5.

II. The Sediment

The 1967 excavations in the cave at Le Lazaret involved a surface of about 55 m². In the upper horizons, near the entrance to the cave, twelve horizons with a total thickness of less than 1 m were identified (Fig. 126). These horizons were often discontinuous.

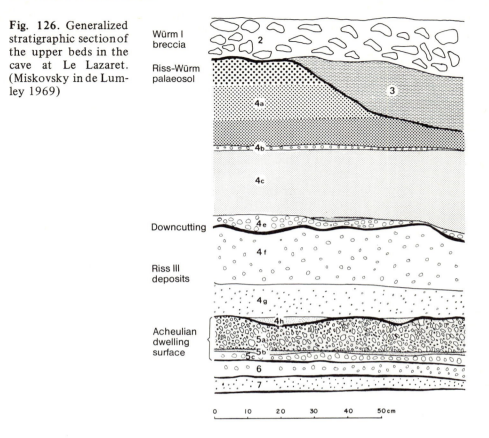

Fig. 126. Generalized stratigraphic section of the upper beds in the cave at Le Lazaret. (Miskovsky in de Lumley 1969)

Würm I breccia

Riss-Würm palaeosol

Downcutting

Riss III deposits

Acheulian dwelling surface

1. The Nature of the Sediment

The sedimentary infill of the cave consisted of brown or yellow sandy clays, with associated pebble beds. The material has a two-fold origin.

The **coarse elements**, boulders, pebbles and coarse sands, come from the disintegration of the roof and walls of the cave under the action of low temperatures. The stones are frequently frost-shattered (Fig. 127). This autochthonous sediment is characteristic of periods of climatic cooling.

The **fine fraction**, with a grain size less than 0.2 mm, was brought into the cave by trickling water either from the entrance or through fissures in the karst. The sediment is generally finely bedded, and is accompanied by small, flat pebbles also formed in the karstic complex. This allochthonous material is more abundant at times of a wet climate.

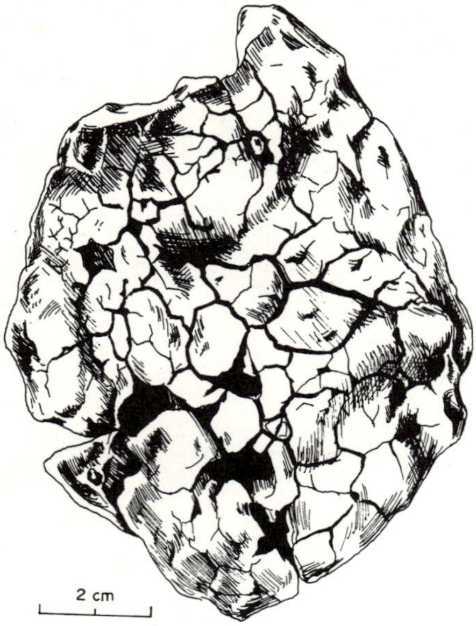

Fig. 127. Frost-shattered stone. (From a photograph by de Lumley 1972)

2 cm

2. The Occupation Level

In the Riss formations (bed. no. 5), an occupation level has been discovered. The sediment there has been considerably disturbed by human activity. The bedding of the fine-grained deposits has been destroyed and the sands are less well sorted than in the other beds. Locally, the horizons have been literally worn away by the coming and going of the people. The small pebbles and frost-shattered stones are less numerous in the living area, while the gravels, which are coarser and more abundant near the entrance to the cave, have doubtless been brought inside by the feet of its occupants. The best proof of human occupation, however, is the high density of implements and bones collected from this horizon.

3. Deposits Post-Dating the Riss Glaciation

The Riss-Würm interglacial is marked by a **palaeosol** containing a horizon of calcareous concretions. This indicates a climatic warming. Above, there is a **stalagmitic floor,** a compact calcareous crust which separates the deposits of the Riss-Würm interglacial from those of the Würm. It suggests a temperate climate which favoured high evaporation of water rich in calcium bicarbonate.

Later, during the Würm, **solifluxion** flows sealed the entrance to the cave.

III. The Fauna and Flora

1. The Fauna

Most of the animals found are still living at the present day.

Level 5, which contains the Acheulian occupation horizon, has been found to be extremely rich in mammal bones, chiefly **deer** (stags, fallow deer) and **caprides** (chamois, ibex), whose incomplete skeletons and bones, which are often reduced to splinters, show that they were brought in by man. **Carnivores** are represented by several wolves and the remains of lynx and panther.

Rabbit bones are abundant, and belong to at least 45 individuals. Scratches from stone implements on the bones indicate that they were hunted by man.

Rodent remains, particularly voles, come from stomach pellets of birds of prey, which probably could not have lived together with man. Marmots, on the other hand, were caught and eaten by the men.

The bones of **birds** are abundant; they belonged to pigeons *(Columba livia)*, rollers *(Coracia graculus)*, blackbirds, wagtails, etc. From their distribution inside the cave, it would seem that they were hunted by man.

Small **sea shells**, unsuitable for food, have been found in the dwelling area. They are littoral grazers (*Littorina, Rissoa,* etc.) and were doubtless brought into the cave with the seaweed which the men used for bedding. Foraminifera have been found in the shells: one, *Globigerina pachyderma,* is a cold water pelagic form.

Various coprolites have been attributed to carnivores and to man.

2. The Flora

The flora contemporaneous with the dwelling horizon is known from pollen analysis of the sediment and from the study of charcoal which came from the occupants' hearths. The dominant species is the *Norway pine*. The pine pollen is sometimes still agglomerated, which means it cannot have travelled far and indicates that the conifers grew close to the cave. Various deciduous (oak) and herbaceous plants (compositaceae, graminaceae) are mixed with the pines.

IV. Human Occupation

Human remains in the cave at Le Lazaret consist of only two teeth and a cranial bone, and are insufficient to reconstruct the physiognomy of

the inhabitants of the cave. On the other hand, man has left many traces
of his stay. These have been discovered in bed 5, which corresponds to
the end of the Riss glaciation.

1. The Dwelling

The implements and animal bones are not scattered at random on the
dwelling surface, but are concentrated in a rectangular area 11 m long by
3.5 m wide (Fig. 128). The perimeter of this surface is marked by blocks
between 5 and 30 cm in diameter. Such a distribution implies the existence of a well-defined space which prevented them from being scattered,
in other words, an **enclosed dwelling**. This construction was held up by
wooden posts supported by stones in a circular pattern. Only these latter
have remained. Probably wooden crosspieces strengthened the verticals.
The whole structure was covered by **skins** which protected the inhabitants from the cold and from the water trickling from the roof of the
cave. A low drystone wall built opposite the entrance to the cave protected the inhabitants from the outside wind.

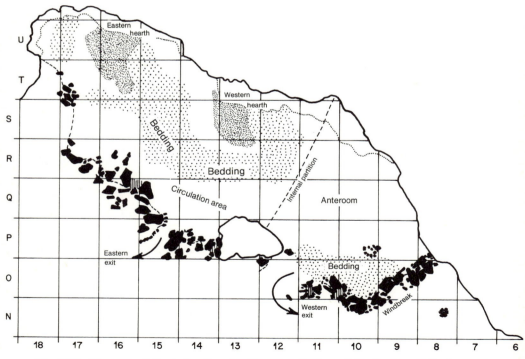

Fig. 128. Diagram of the Acheulian dwelling area (bed no. 5) in the cave at Le Lazaret.
(De Lumley 1969)

The belt of boulders which must have anchored the bottoms of the skins to the ground is broken in two places which doubtless reflects the positions of the **entrances**. The main opening faced towards the back of the cave. Probably an internal partition of skins divided the dwelling into two compartments of unequal size.

Inside the dwelling area, two small **hearths**, dug into the clay of the floor and probably used for heating, have been found. The large fires needed for cooking food must have been outside. Bones and implements are more numerous close to the hearths. A **circulation area** has been identified, where objects have been pushed aside and roughly aligned by the feet of the men. The presence of small marine shells, which are normally attached to plants in the littoral zone, indicates the position of the **bedding**: they are mainly concentrated around the hearths. Their distribution coincides with that of the distal bones from the feet of furry animals (carnivores) which suggests that the bedding was covered by skins which had been crudely prepared, retaining the feet.

To summarize, the dwelling of these Acheulian men consisted of a large **tent** erected inside the cave (Fig. 130).

2. The Implements

The men who lived in the cave at Le Lazaret abandoned many stone and a few **bone implements** on the ground (Fig. 129).

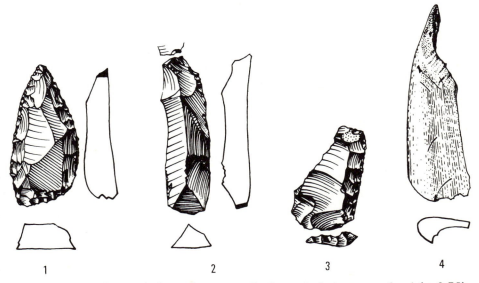

Fig. 129. Acheulian tools from the cave at Le Lazaret. *1* stone spearhead (× 0.75); *2* stone blade (× 0.75); *3* stone scraper (× 0.75); *4* bone piercer (× 0.75). (De Lumley 1969)

The stone was worked by hitting boulders or pebbles of hard rock (flint, sandstone, limestone), collected from the alluvium of nearby rivers, together. The flakes were then trimmed and used as scrapers, chisels, knives, etc. The typological and technical features of this industry belong to the late Acheulian.

3. Human Activity

The main occupation of the inhabitants of the cave at Le Lazaret was **hunting**. From the bones found at the tent locality, the game was varied: stags, fallow deer, ibex, chamois, lynx, wolves, foxes, marmots, etc. The men used stone-pointed weapons but probably caught the small mammals and birds in **traps**. Hunting large cats like the panther needed the co-operation of several men and implies some **social organisation**.

The meat was butchered and cooked outside the cave. The men prepared the **skins** in the shelter of the tent, doubtless using the many bone splinters for this. To sew the skins together or to make traps, they used mammal tendons. The striations seen on the tibia of many of the rabbits were made during the removal of tendons using stone implements.

The **knapping** of the hard rocks must have been carried out outside the cave by daylight. Only the finishing of the implements took place inside the tent.

4. The Length of Occupation

The length of occupation of the site can be worked out with some accuracy. Amongst the remains of meals in the lower horizons of bed 5 are young ibex about 5 months old. Ibex are generally born about the middle of June. This suggests that the men sought the shelter of the cave during November in order to spend the winter there. The upper horizons of bed 5 contain the remains of marmots killed by man. These animals wake from hibernation and leave their burrows at the beginning of spring. Shortly afterwards, possibly in mid-April, man finally left the cave. Their short 5-month stay there allowed them to shelter from the bad weather before recommencing their nomadic life. As they left, the hunters placed the skull of a wolf near the entrance to the cave; possibly this had some magical significance. The remains of the mens' meals in turn attracted a crowd of rodents and carnivores who left their teeth marks on the bones. Then, in their turn, the birds of prey, recognisable from their stomach pellets, regained possession of the cave.

V. The Climate

The relative abundance of **frost-shattered pebbles** in the late Riss sediments indicates three short periods of climatic cooling, the worst coinciding with the stay of man. However, winter temperatures today inside the cave are always above those of the outside air. For frost to shatter rocks within the cave, the winter temperatures must have been very low, lower than any on the Mediterranean coast today. However, the occurrence at every horizon of fine sediment brought in by trickles of water suggests that the climate was always wet.

Most of the animal and plant species collected at the level of the dwelling surface also indicate a climate which was noticeably colder than that at the present day in the Nice region. This is true especially of the chamois, ibex, marmots and wolves and, amongst the plants, the Norway pine. However, the discovery of some bactracian bones precludes very great cold.

O_{16}/O_{18} isotope analysis on the carbonate of the marine shells confirms this conclusion: the temperature of the sea was on average 6°C colder than at the present day.

To summarise: towards the end of the Riss glaciation, the climate in the Nice region was cold temperate. It was wetter in autumn and winter when temperatures dropped well below 0°C. The summers, on the other hand, must have been warm and sunny.

Fig. 130. Reconstruction of the Acheulian hut in the cave at Le Lazaret. (De Lumley in Chaline 1972)

VI. Conclusion

Towards the end of the Riss glaciation, 130,000 years ago, a group of
Acheulian men, family or tribe, sought shelter in the cave at Le Lazaret,
which provided both protection against the cold and a viewpoint to see
along the shore. They built a tent for rest, covered by animal skins, and
organised hearths and bedding there (Fig. 130). At the time, the slopes
of Mont Boron were covered by Norway pines while further away was
a mixed oak forest where stags and fallow deer lived. Rabbits were abundant in the undergrowth. Higher up, they found ibex, chamois and marmots. The hunters thus found their food close to home.

They arrived at the beginning of a hard winter when the outside temperature dropped well below freezing and then took up their nomadic
life again when the good weather came.

Man appeared several millions of years ago. By means of technology,
he has extended his grip on the world, changed it and turned to his own
profit, or sometimes his loss, the natural processes of sedimentation and
the activity of living organisms. From this point on, it is the turn of the
historian to continue the tale.

Appendix I Guide to the Palaeoecological Study of Fossiliferous Horizons

The Fossils

1. Make a qualitative and quantitative list, unit by unit, of the macrofossils, microfossils and evidence of biological activity.
2. Study the distribution of the organisms in the sediment:
   - orientation (fossils in life position, current alignment, lying oblique to bedding, etc.);
   - density (lenses, concretions, etc.);
   - sorting (frequency curves);
   - state of preservation (disarticulation of hard parts, abrasion, comminution, etc.);
   - evidence of reworking (geopetal infillings, mixing of faunas and floras).
3. Note the adaptive features of the organisms in relation to the physico-chemical conditions (substrate type, salinity, water turbulence, etc.).
4. Note the associations of organisms (epizoans, epiphytes, etc.).
5. Note the features of fossilisation (preservation of soft parts, solution, mineralisation, etc.).

The Sediment

1. Determine the lithology and grain size of the surrounding sediments.
2. Observe the nature of the bedding.
3. List the sedimentary structures (ripple marks, flute casts, synsedimentary folding, etc.) and note their position within the bed.
4. Identify the possible palaeocurrent directions.
5. Make a diagram of the sedimentary succession and the distribution of the fossils within it.

Interpretation of the Sedimentary and Biological Environments

1. Define the environment (sea, lake, fluviatile complex, etc.).
2. Specify the physico-chemical characteristics of the environment (water depth, energy level, rate of sedimentation, climate, etc.).
3. Reconstruct the animal and plant populations.
4. Analyse the diagenetic history.

Appendix II Table Summarizing the Principal Features of the Environment

	Environmental features	Sediment	Fossils
Salinity	Marine	Minerals: glauconite, chamosite Clay geochemistry: boron > 100 ppm Rocks: abundant carbonates	Stenohaline fauna, high diversity
	Brackish or hypersaline	Minerals: evaporites	Euryhaline fauna, low diversity
	Freshwater	Clay geochemistry: boron < 100 ppm Fluviatile environment: unimodal currents	Freshwater fauna and flora
Depth	Deep water	Fine-grained sediments, turbidites	Plankton, carnivores, feeding trails
	Shallow water	Coarse sediments (sands, oolites, etc.), oscillation ripples, crescent marks, flaser bedding, curved cross-bedding	Green plants, herbivorous fauna, dwelling traces, resting traces, locomotion traces, (tetrapod, vertebrates, birds, etc.)
	Intertidal zone	Dolomite frequent, beach sands, birdseye texture, desiccation cracks, bimodal palaeocurrents	Algal structures (stromatolites), fixed or endobiontic fauna (vertical or oblique burrows)
	Continental environment	Palaeosols, desiccation cracks	Terrestrial fauna and flora, eggs, roots in situ

	Environmental features	Sediment	Fossils
Water turbulence	Active water	Rounded grains, sparite cement grading, scour and current marks, cross-bedding	Suspension feeders, massive forms, fossils disarticulated, worn, sorted, oriented
	Quiet water	Clay-micrite, horizontal bedding	Detritus feeders, faecal pellets, branching forms, well-preserved fossils
Oxygenation of the water	Oxygenated environment	Minerals: ferric oxides, red colouration common	Abundant benthos and endofauna
	Anaerobic environment	Minerals: pyrite, siderite, barely decomposed organic matter	Benthos and endofauna absent, fossils well-preserved
Temperature	Hot climate	On land: intense alteration. In water: limestones and evaporites common	Thick calcareous shells, reefs
	Cold climate	Glacial erosion, periglacial phenomena	Thin calcareous shells, siliceous organisms common
Substrate	Indurated	Mineralised surfaces	Boring and encrusting organisms
	Mobile		Burrowing organisms
Rate of Sedimentation	High		Fossils lying oblique to bedding Organisms well-preserved
	Low	Hardgrounds	Altered fossils

Appendix III   Stratigraphic Position of Some European Fossiliferous Horizons

The figures in millions of years indicate the beginning of the relevant epoch

	Periods		Continental horizons	Marine horizons
Quaternary Era — 2 MA			Dordogne (Lascaux) Languedoc (Hortus) Alpes maritimes (Le Lazaret) Dordogne (Les Eyzies) Vallée du Rhône (St. Vallier)	
Tertiary Era	Pliocene — 5 MA		Harz (Willershausen)	
	Miocene — 25 MA		Greece (Pikermi) Lake Constance (Öhningen)	Touraine: shell beds
	Oligocene — 37 MA		Quercy: phosphates Baltic Amber Aix-en-Provence	
	Eocene — 65 MA	Upper	Gard (Robiac)	Paris Basin: shell beds, gypsum
		Middle	Alsace (Bouxwiller) Hesse (Messel) Saxe (Greiseltal)	
		Lower	Marne (Sézanne)	
Mesozoic Era	Cretaceous — 140 MA	Upper		Rudist reefs: Aquitaine, Corbières (montagne des Cornes) Provence (Martigues)
		Lower	Hainaut (Bernissart)	
	Jurassic — 200 MA	Upper	Portugal (Guimarota)	French Jura (Cerin) Bavaria (Solnhofen) Yonne (Oxfordian reefs) Spain (Sierra de Montsech)
		Middle		Moselle
		Lower		Württemberg (Holzmaden)

	Periods		Continental horizons	Marine horizons
Mesozoic Era	Trias − 240 MA	Upper		Eastern Alps (Hoher Göll)
		Middle		Catalonia (Montreal) Tessin (Monte San Giorgio)
		Lower	Vosges (Grès à Voltzia)	
Palaeozoic Era	Permian − 280 MA		Massif Central (Autun)	
	Carboniferous − 345 MA		Massif Central Sarre, Franco-belgian basin (coal basins)	
	Devonian − 395 MA	Upper	Scotland (Achanarras)	Ardennes (Frasnian reefs)
		Lower	Scotland (Rhynie)	Hunsrück
	Silurian − 430 MA			Sweden (Gotland reefs)
	Ordovician ? − 500 MA			Anjou (Angers) Norway (Oslo)
	Cambrian − 570 MA			Bohemia (Jince)

References

Ager DV (1963) Principles of paleoecology. An introduction to the study of how and where animals and plants lived in the past. McGraw-Hill, New York

Aldinger H (1965) Zur Ökologie und Stratinomie der Fische des Posidonienschiefers (Lias epsilon). Senckenberg Lethaea 46A:1–12

Allen JRL (1965) A review of the origin and characteristics of recent alluvial sediments. Sedimentology 5:89–191

Allen JRL (1970) Physical processes of sedimentation. An indroduction. Allen and Unwin, London

Allen JRL (1973) Compressional structures (patterned ground) in Devonian pedogenic limestones. Nature Phys Sci 243:84–86

Allen JRL, Tarlo LB (1963) The Downtonian and Dittonian facies of the Welsh Borderland. Geol Mag 100:129–155

Alpern B et al (1970) Pétrologie des charbons. Ann Soc Géol Nord 90:203–222

Aubouin J, Brousse R, Lehman JP (1968) Précis de Géologie, Tome I: Pétrologie, 712 p; Tome III: Téctonique Morphologie Globe terrestre, 549 p. Dunod, Paris

Babin C (1971) Eléments de paléontologie. Colin, Paris

Barrabe L, Feys R (1965) Géologie du charbon et des bassins houillers. Masson, Paris

Barthel KW (1970) On the deposition of the Solnhofen lithographic limestone (Lower Tithonian, Bavaria, Germany). N Jahrb Geol Paläont Abh 135:1–18

Barthel KW (1978) Solnhofen. Ein Blick in die Erdgeschichte. Ott, Thun

Bellair P, Pomerol C (1965) Eléments de Géologie. Colin, Paris

Bersier A (1958) Séquences détritiques et divagations fluviatiles. Eclogae Geol Helv 51:854–893

Brenner K (1976) Biostratonomische Untersuchungen im Posidonienschiefer (Lias epsilon, Unteres Toarcium) von Holzmaden (Württemberg, Süd-Deutschland). Zbl Geol Paläont II:223–226

Brenner K, Seilacher A (1978) New aspects about the origin of the Toarcian Posidonia Shales. N Jahrb Geol Paläont Abh 157:11–18

Buisonje PH de (1972) Recurrent red tides, a possible origin of the Solnhofen Limestone. Proc Kkl Nederl Akad Wetensch B 75:152–177

Burrolet PF, Byramjee R, Couppey C (1966) Contribution à l'étude sédimentologique des terrains dévoniens du nord-est de l'Ecosse. Notes et Mémoires Comp Fr Pétroles, vol 9, Paris

Busson G (1968) La sédimentation des évaporites. Comparaison des données sahariennes à quelques théories, hypothèses et observations classiques ou nouvelles. Mem Mus Nat Hist Nat Sér C/19:125–169

Chaline J (1972) Le Quaternaire. L'histoire humaine dans son environnement. Doin, Paris

Dechaseaux C (1951) Contribution à la connaissance des estéries fossiles. Ann Paléont 37:123–132

Duff PMD, Hallam A, Walton EK (1967) Cyclic sedimentation. Elsevier, Amsterdam

Einsele G, Mosebach R (1955) Zur Petrographie, Fossilerhaltung und Entstehung der Gesteine des Posidonienschiefers im Schwäbischen Jura. N Jahrb Geol Paläont Abh 101:319–430

Fayol H (1888) Résumé de la théorie des deltas et histoire de la formation du bassin de Commentry. Bull Soc Géol France 16:968–979

Fischer JC (1970) Tendances et méthodes en paléoécologie. Bull Soc Géol France 12: 318–326

Gall JC (1971) Faunes et paysages du Grès à Voltzia du Nord des Vosges. Essai paléo-écologique sur le Buntsandstein supérieur. Mém Serv Carte Géol Als Lorr No 34. Strasbourg

Gall JC (1971) Le Grès à Voltzia des Vosges: le passage d'un paysage deltaïque à un environnement littoral. CR Acad Sci 273D:2449–2452

Gall JC (1972) Le rôle de la paléoécologie dans la reconstitution des anciens biotopes. Application aux gisements fossilifères du Trias inférieur des Vosges. Bull Soc Ecologie 3:354–367

Gall JC, Grauvogel L (1966) Pontes d'invertébrés du Buntsandstein supérieur. Ann Paléont (Invertébrés) 52:155–161

Glaessner MP, Wade M (1966) The late Precambrian fossils from Ediacara, South Australia. Palaeontology 9:599–628

Goldring R, Curnow CN (1967) The stratigraphy and facies of the late Precambrian at Ediacara, South Australia. J Geol Soc Australia 14:195–214

Goldring R, Seilacher A (1971) Limulid undertracks and their sedimentological implications. N Jahrb Geol Paläont Abh 137:422–442

Grauvogel L (1947) Contribution à l'étude du Grès à Voltzia. CR Somm Soc Géol France 25–37

Grauvogel-Stamm L (1969) Nouveaux types d'organes reproducteurs mâles de conifères du Grès à Voltzia (Trias inférieur) des Vosges. Bull Serv Carte Géol Als Lorr 22:93–120

Guber Y et al (1966) Essai de nomenclature et charactérisation des principales structures sédimentaires. Technip, Paris

Hallam A (1961) Origin des cycles mineurs de sédimentation carbonatée dans le Lias. Mem BRGM 4:171–175

Häntzschel W (1962) Trace fossils and problematica. In: Moore RC (ed) Treatise on invertebrate paleontology, part W. Geol Soc Am, Boulder, p 177

Häntzschel W, El-Baz F, Amstutz GC (1968) Coprolites. An annotated bibliography. Geol Soc Am Mem 108. Boulder

Hauff B (1960) Das Holzmadenbuch. Rau, Öhringen

Heckel PH (1972) Recognition of ancient shallow marine environments. In: Rigby JK, Hamblin WK (eds) Soc Econ Paleont Miner Spec Publ 16, Tulsa

Hecker RF (1957) Bases de la Paléoécologie, traduit par Roger J en 1960. Bur Rech Géol Min Ann Inf Géol 44, Paris

Imbrie J, Newell N (1964) Approaches to paleoecology. Wiley, New York

Janicke V (1969) Untersuchungen über den Biotop der Solnhofener Plattenkalke. Mitt Bayer Staatssamml Paläont Hist Geol 9:117–181

Kauffman EG (1978) Benthic environments and palaeoecology of the Posidonien-schiefer (Toarcian). N Jahrb Geol Paläont Abh 157:18–36

Keupp H (1977) Ultrafazies and Genese der Solnhofner Plattenkalke (Oberer Malm, Südliche Frankenalb). Abh Naturhist Ges Nürnberg 37:128

Kuhn O (1961) Die Tier- und Pflanzenwelt des Solnhofener Schiefers. Geologica Bavarica 44:68

Ladd HS (1957) Treatise on marine ecology and palaeoecology, vol 2: Palaeoecology. Geol Soc Amer Mem 67

Laporte LF (1968) Ancient environments. Prentice-Hall, New Jersey

Lehman JP (1967) La paléoécologie. Mises à jour scient 1:369–385

Lessertisseur J (1955) Traces fossiles d'activité animale et leur signification paléobiologique. Mém Soc Géol France N S 74

Lombard A (1972) Séries sédimentaires. Genèse – évolution. Masson, Paris

Lumley H de et al (1969) Une cabanne acheuléenne dans la grotte du Lazaret (Nice). Mém Soc Préhist Fr 7

Lumley H de et al (1972) La grotte de l'Hortus (Valfalunès, Hérault). Les chasseurs néandertaliens et leur milieu de vie. Elaboration d'une chronologie du Würmien II dans le Midi méditerranéen. Etudes Quatern Mém 1

Menot JC, Rat P (1967) Sur la structure du complex récifal jurassique supérieur de la vallée de l'Yonne. CR Acad Sci 264D:2620−2623

Millot G (1964) Géologie des argiles. Altérations, sédimentologie, géochimie. Masson, Paris

Müller AH (1963) Lehrbuch der Paläozoologie. I. Allgemeine Grundlagen. Fischer, Jena

Perès JM (1965) Réflexions sur les rapports entre l'écologie et la paléoécologie marines. Palaeogeogr Palaeoclimatol Palaeoecol 1:51−68

Perreau M, Pomerol C (1969) Etude paléontologique et paléoécologique du rivage bartonien d'Attainville (Val d'Oise). Bull Soc Géol France 11:13−27

Perreau M, Pomerol C (1971) Répartition des testes d'organismes dans les chenaux, dans l'Auversien du Plessis-Gassot (Val d'Oise). CR Somm Soc Géol France 367−369

Pettijohn FJ, Potter PE (1964) Atlas and glossary of primary sedimentary structures. Springer, Berlin Göttingen Heidelberg

Piveteau J (1951) Images des mondes disparus. Masson, Paris

Piveteau J (1973) Origine et destinée de l'Homme. Masson, Paris

Plaziat JC (1971) Racines ou terriers? Critère de distinction à partir de quelques exemples du Tertiaire continental et littoral du Bassin de Paris et du Midi de la France. Conséquences paléogéographiques. Bull Soc Géol France 13:195−203

Pomerol C (1964) Origine et conditions de sédimentation des dépôts sableux et argileux dans le golfe bartonien du bassin de Paris. In: Van Straaten MJU (ed) Deltaic and shallow marine deposits. Elsevier, Amsterdam

Pomerol C (1965) Les sables de l'Eocène supérieur (Lédien et Bartonien) des bassins de Paris et de Bruxelles. Mém Expl Carte Geol Dét France, Paris, 214 p

Pomerol C (1973) Stratigraphie et paléogéographie: ère cénozoïque (Tertiaire et Quaternaire). Doin, Paris

Pomerol C, Feugueur L (1968) Bassin de Paris. Ile-de-France. Guides géologiques régionaux. Masson, Paris

Pomerol C, Trichet J (1969) Présence des grès de plages dans l'Auversien et le Marinésien du Bassin de Paris. CR Somm Soc Géol France 129−130

Rayner DH (1963) The Achanarras Limestone of the Middle Old Red Sandstone, Caithness, Scotland. Proc Yorks Geol Soc 34:117−138

Reineck HE, Singh IB (1973) Depositional sedimentary environments. Springer, Berlin Heidelberg New York

Roger J (1974) Paléontologie générale. Masson, Paris

Sacchi CF, Testard P (1971) Ecologie animale. Organismes et Milieux. Doin, Paris

Schäfer W (1962) Aktuo-Paläontologie nach Studien in der Nordsee. Kramer, Frankfurt am Main

Seilacher A (1953) Die geologische Bedeutung fossiler Lebensspuren. Z Dtsch Geol Ges 105:214−227

Seilacher A (1953) Studien zur Palichnologie. I. Über die Methoden der Palichnologie. N Jahrb Geol Paläont Abh 96:421−452

Seilacher A (1954) Studien zur Palichnologie. II. Die fossilen Ruhespuren (Cubichnia). N Jahrb Geol Paläont Abh 98:87−124

Seilacher A (1967) Bathymetry of trace fossils. Marine Geology 5:413−428

Seilacher A (1970) Begriff und Bedeutung der Fossil-Lagerstätten. N Jahrb Geol Paläont Mh 1:34−39

Seilacher A (1971) Preservational history of ceratite shells. Palaeontology 14:16−21

Selley RC (1970) Ancient sedimentary environments. Chapman and Hall, London

Tasch P (1957) Flora and fauna of the Rhynie Chert: a palaeoecological reevaluation of the published evidence. Bull Univ Wichita 32:24

Termier H, Termier G (1968) Biologie et écologie des premiers fossiles. Masson, Paris

Thaler L (1965) Les œufs du Dinosaures du Midi de la France livrent le sécret de leur extinction. Science-Progrès-La Nature 3358:41–48

Theobald N (1972) Fondements géologiques de la Préhistoire. Essai de chronostratigraphie des formations quaternaire. Doin, Paris

Van Straaten LMJU (1971) Origin of Solnhofen Limestone. Geol Mijnbouw 50:3–8

Vatan A (1967) Manuel de sédimentologie. Technip, Paris

Vetter P (1968) Géologie et paléontologie des bassins houillers de Decazeville, de Figeac et du détroit de Rodez, vol 1 and 2. Hoillères du Bassin d'Aquitaine, Albi

Wade M (1968) Preservation of soft-bodied animals in Precambrian sandstones at Ediacara, South Australia. Lethaia 1:238–267

Zankl H (1967) Die Karbonatsedimente der Obertrias in den nördlichen Kalkalpen. Geol Rundschau 56:128–139

Zankl H (1969) Der Hohe Göll. Aufbau und Lebensbild eines Dachsteinkalk-Riffes in der Obertrias der nördlichen Kalkalpen. Abh Senckenberg Naturforsch Ges 519:123

Zankl H (1971) Upper Triassic carbonate facies in the Northern Limestone Alps. Sedimentology of parts of Central Europe. Intern Sediment Congress Guidebook VIII:147–185

Zeigler B (1972) Allgemeine Paläontologie. Schweizerbart, Stuttgart

Subject Index

Abyssal zone 82
   fauna in 27
acanthodians 117, 120, 131
Acanthoteuthis 170
Achanarras 119
Acheulian 186–194
acritarchs 8
actinopterigians 120
active water 102
adaptation to turbulence 25
Aeger 169
aeolian dunes 60, 180
   environment 52
   sedimentation 81
Aethereophyllum 140
agitation of water 24
agnathans 117, 121
alcyonarians 112
Alethopteris 131
algae 9, 83
   adptation to turbulence 25
   blue-green 23, 27, 86, 174, 176
   brown 26, 86
   calcareous 86, 153
   dasyclad 181
   depth distribution of 27
   encrusting 86
   green 8, 23, 26, 27, 154, 181
   Jurassic 27
   Precambrian 111, 112
   red 23, 26, 27, 154
   salinity distribution of 23
   sargassid 8
   stromatolites 22, 59, 86
   symbiotic 86
algal blooms 96, 122
   hydrocarbons 8
   mats 149, 151, 156, 157, 164, 174
allogromids, depth zonation of 27
Allohippus 31
alluvial plain 119
alteration 49
alternating sequences 71
alternation of generations 36
amber 97, 99

ambulacra 6, 13
ammodiscids 137
ammonites, centre of buoyancy 19
   Liassic 159
   Portlandian 170
amphibians 15
   freshwater 24
   Gres a Voltzia 140, 145
amphipods 13
anaerobic bacteria 133
   environments 42, 72, 99, 121, 132,
     163, 165, 174
anhydrite 94
annelids 4, 5, 9, 14, 15, 138
   borings 41, 154
   burrows 41, 42
   depth zonation of 27
   Eocene 182
   freshwater 24
   Precambrian 111, 112
   salinity distribution of 23
Anomopteris 140
Anthracomya 131
antidunes 67
anti-ripplets 120, 121
Antrimpos 138, 139
aphroditids 138
Aporrhais 5
aptychus 159, 170, 173
aquatic environments, substrate of 20
   temperature 32
Aquitaine Basin 48
Aquitainian 73
Arcestes 150
archaeocyathids 4, 13, 22, 86
Archaeopteryx 172
Ardennes, Devonian reefs 25
   pebbles from, in Eocene 179
Areinicola 42
arenicolids 13
arkose 49
arthropods 10, 14
   buccal appendages 15
   burrows 41
   Grès à Voltzia 138

Fossil Algae

Recent Results and Developments

Editor: **E. Flügel**

1977. 119 figures, 32 plates.
XI, 375 pages
ISBN 3-540-07974-2

Contents: Blue-Green Algae and Stromatolites. – Green Algae. – Red Algae. – Problems of Affinities. – Biometry. – Ultrastructure. – Algae and Sedimentary Environments.

This collective work results from the 1st International Symposium on Fossil Algae at the Institute of Paleontology of the University of Erlangen-Nürnberg in October 1975, and presents a broad coverage of present research. The main emphasis of the book lies on critical studies of the phylonge etical stratigraphical and sedimentological significance of blue-green algae and stromatolites, dasyclads, and of Paleozoic and Mesozoic red algae. Problems of the affinities of calcerous algae and "pseudo-algae" are discussed as well as the ultra-structure of carbonate producing algae. The cooperation of more than 50 geo- and bioscientists has provided a wide range of information on a field of research equally important for paleontologists, geologists, and biologists.

Springer-Verlag
Berlin
Heidelberg
New York
Tokyo

Cyclic and Event Stratification

Editors: **G. Einsele, A. Seilacher**

1982. 180 figures. XIV, 536 pages
ISBN 3-540-11373-8

Contents: Limestone–Marl Rhythms and Climate-controlled Facies Changes. – Event Stratification: Calcereous and Quartz-Sandy Tempestites. Other Event Deposits. – Cyclicity and Event Stratification in Black Shales. – Summary.

Stratifiation is a fundamental property of sedimentary rocks which is used in stratigraphic correlation, structural and economic geology. This book focuses on two major processes that produce pronounced bedding in the marine realm:

– cyclic or periodic gradual changes of environmental conditions, and

– rare and unpredictable events (storm wave-induced tempestites; turbidites).

These processes affect a great variety of sediments including black shales and carbonates, where primary features can be masked or overprinted by later diagensis. Periodic phenomena with potential time information are represented by some limestone-marl successions.
The 44 contributions by authors from many countries range from detailed descriptions of characteristic stratification phenomena, sedimentary and biostratinomic structures and faunal responses to articles presenting general conceptual models and interpretations of the depositional environment. Unique in its summary of the state-of-the-art in this interdisciplinary field, **Cyclic and Event Stratification** is of interest to all those concerned with bedded sequences.

Springer-Verlag
Berlin
Heidelberg
New York
Tokyo